Will We Ever Speak Dolphin?

and 130 more science questions answered

NewScientist

Will We Ever Speak Dolphin?

and 130 more science questions answered

edited by Mick O'Hare

Nicholas Brealey Publishing
London • Boston

First published in Great Britain in 2012 by Profile Books Ltd

This edition published in 2019 by Nicholas Brealey Publishing, an imprint of John Murray (Publishers)

1

A CIP catalogue record for this title is available from the British Library and the Library of Congress

ISBN 978-1-5293-0935-5
Ebook ISBN 978-1-5293-3711-2

Text design by Sue Lamble
Typeset in Palatino by MacGuru Ltd

Printed and bound in the United States of America.

John Murray policy is to use papers that are natural, renewable and recyclable products and made from wood grown in sustainable forests. The logging and manufacturing processes are expected to conform to the environmental regulations of the country of origin.

Nicholas Brealey Publishing
Hachette Book Group
53 State Street
Boston, MA 02109

www.nbuspublishing.com

Contents

Introduction

Ian Fleming, the author of the James Bond spy novels, really didn't know what he had started when he decided the fictional spy should order his vodka martinis shaken, not stirred. First it became one of the most famous catch-phrases in movie history. Now it's become the subject of an entire chapter in this latest edition of science questions and answers from *New Scientist*. Just why did Bond want that martini stirred? The debates have raged long and hard down the years but now we think we've cracked it. Turn to page 88 for the full lowdown on the science of Bond's iconic tipple. We think there's very little more to learn now, but conceit is the downfall of any scientist, so if anybody out there knows better our contact details are below.

And not only did James Bond enjoy his martinis, he also had – according to the screenplay of *Thunderball* – a double first in Oriental Languages from Oxford University. So if he hadn't been a fictional character, he'd have been just the person to answer the title question of this book (see p. 85). As an accomplished linguist, if Dolphin could be learned, he'd have learned it – not least because it would have come in handy in *The Spy Who Loved Me*. It's surely not improbable that undersea megalomaniac and archetypal Bond villain Karl Sigmund Stromberg had been learning Dolphin as part of his evil plan to relocate the human species underwater, so Bond too would surely have swotted up to ensure he could foil yet another madman. Unlikely, you say? Well, it's perhaps not as improbable as the fact that in more than twenty movies

Bond has been shot at more times than any other fictional hero, yet never taken that final, fatal bullet. Check out just how improbable on p. 102.

Of course, there's more to this book than a fictional spy and his foibles. Do you know why we become hoarse when we shout, whether it's better for the planet if we all become vegetarians, or why we want to urinate more in cold weather? Well we didn't either until somebody bothered to get in touch and ask us and then somebody else gave us the answer. The Last Word column in *New Scientist* magazine – which gave birth to this book and its bestselling predecessors including *Does Anything Eat Wasps?* and *Why Don't Penguins' Feet Freeze?* – has been answering everyday science questions from the general public since 1994. You can ask one yourself or, even better, answer one by buying the weekly magazine or visiting us online at www.newscientist.com/lastword.

This volume, of course, contains all new material save for a small number of the answers at the start of Chapter 4 which provide background science to the vodka martini story. Now grab your vodka, your vermouth and your olive jar and read on. And after you've shaken or stirred do let us know if you disagree with our cocktail of conclusions. There may still be another chapter waiting to be written on Bond's favourite drink.

Mick O'Hare

Special thanks – in no particular order – to the following for their help in producing this book: Richard Fisher, Melanie Green, Eleanor Harris, Judith Hurrell, the *New Scientist* subs and art teams, Jeremy Webb, Beverley De Valmency, Paul Forty, Valentina Zanca, Andrew Franklin, Stephanie Pain, Ben Longstaff, Runyararo Chivaura, Dirk Kuyt, Nick Heidfeld, Sally and Thomas.

1 Food and drink

Squeaky cheese

Why does halloumi cheese squeak against your teeth as you eat it?

Nikos Skouris
Nicosia, Cyprus

This is an example of the stick-slip phenomenon. The cheese is rubbery and as your teeth begin to squeeze it, the halloumi deforms with increasing resistance until it loses its grip and snaps back to something like its original shape. At the point where the slipping stops it regains its grip and the process repeats, commonly at a frequency near 1000 hertz, give or take an octave or two. The vibration produces a squeal of corresponding frequencies that may vary with the circumstances, such as whether the cheese has oil on it.

Squeaky halloumi is enough to make some people's toes curl, like fingernails dragged down a blackboard. This is because such sounds often warn of injury – a broken bone grating – or an unpleasant sensation, such as sand in your teeth, or stone abrading fingernails.

Probably long before our ancestors evolved into apes, they developed an inherited distaste for such noises and the associated sensations. It was likely an evolutionary adaptation to their way of life; those who did not respond to the signals tended to have shorter and less productive lifespans.

Jon Richfield
Somerset West, South Africa

Cereal cement

My two favourite breakfast cereals are Shreddies and Weetabix. When I've finished, the remnants in the bowl look similar, but I can always tell which was which: a Shreddies bowl can be washed up quite easily, while Weetabix clings like cement. Why the difference?

Frank Johnson
Birmingham, UK

As a lifelong consumer of Weetabix, I feel qualified to answer this question. Both Shreddies and Weetabix contain a high proportion of starch, which can form an adhesive paste with water. This phenomenon is well known to bookbinders because it is used to make paper.

Starch consists of a mixture of amylose and amylopectin, polymers that can absorb water to form a gel. As the gel dries, the water is expelled and bonds between the molecular chains reform, creating a semi-crystalline 'cement' which will adhere to any adjacent surface.

This effect is much more noticeable with Weetabix than Shreddies simply because Weetabix is made of fine flakes of cereal compressed together, while Shreddies are made from longer strands. That means Weetabix has a greater surface area of adhesive in contact with the bowl, making it more difficult to clean.

Chris Sugden
London, UK

I have no experience of Shreddies, but am familiar with the Weetabix problem. So while I don't know the difference between the properties of the two cereals, I can give this advice to your correspondent. Soak the used Weetabix bowl for a few minutes, rather than a few seconds, before cleaning – it makes it much easier.

David Purchase
Bristol, UK

A spoonful of sugar...

If I leave my jar of brown sugar standing overnight, the surface crystals will bind together and I will need a spoon to scrape and loosen them so I can pour the sugar out. What property of brown sugar causes the surface to bind together so quickly?

Peter Franks
Sydney, Australia

This question was really nostalgic for me. Many years ago I collaborated with David Bagster, a chemical engineer at the University of Sydney, whose research career was dedicated to the wayward properties of unrefined sugar and how to overcome them to allow it to be handled in bulk.

Raw or brown sugar crystals have a permanent liquid layer on their surface. Normally this is apparent only as an obvious stickiness but if it can evaporate, as in your correspondent's sugar jar, the sugar in the surface layer will crystallise and cement the grains together. In cold weather the sugar can crystallise throughout the whole mass, turning it solid.

Bagster told me of a spectacular case of this, in which a bulk transport ship took on a load of raw sugar in the tropics destined for a cold-water Russian port. On arrival the sugar had set like rock and was immovable. The last I heard, the situation hadn't been resolved and the ship was still clinging on to its load.

Guy Cox
Australian Centre for Microscopy and Microanalysis
University of Sydney
New South Wales, Australia

If the humidity around brown sugar is low then it will dry and clump into hard pieces, making it difficult to pour. This happens to my brown sugar even in a jar with a lid.

To prevent this, during the winter months I put a piece of bread or apple peel into the container. The sugar then stays moist and free from clumping.

Gina Kirby
New Maryland, New Brunswick, Canada

Those garlic blues

I made a salad dressing with olive oil, apple cider vinegar, garlic cloves chopped into halves, fresh ginger, mixed herbs and mustard powder. When the dressing was finished I put a lidded jar of it in the fridge, and two days later topped it up again with fresh ingredients. The following day the garlic from the original batch had turned bright blue. Why?

Ellice Bourke
Katherine, Northern Territory, Australia

The discolouration is the result of some complicated chemistry involving the garlic's flavour compounds. The phenomenon is confusingly called 'greening', and the food industry has encountered enough accidentally coloured batches of processed garlic for it to have generated some interest.

In the traditional Chinese pickle of garlic cloves in vinegar known as Laba garlic, the colouration is intentional. Chemists have speculated on its cause since at least the 1940s, and in the last few years Chinese and Japanese researchers have worked out what is going on.

The flavour of garlic is generated when an enzyme called alliinase acts on stable, odourless precursors. These are normally in separate compartments in the cell but can combine if there is damage, including that caused by vinegar. The major flavour precursor in garlic is alliin (S-2-propenyl cysteine sulphoxide) while a minor one is isoalliin ((E)-S-1-propenyl cysteine sulphoxide).

Key to the colour change is a product of these reactions called di-1-propenyl thiosulphinate. It can react at slightly acid pH with amino acids from the ruptured cells to form pyrrole compounds, which are then linked together by di-2-propenyl thiosulphinates to form dipyrroles. These are reddish purple, but as the cross-linking continues, molecules with deeper and bluer hues are formed. Among these are compounds called phycocyanins, which are related to chlorophylls and are found in some algae that are used as blue colouring by the food industry.

Keeping garlic somewhere cool increases the amount of isoalliin present, which is why the best Laba garlic is produced several months after harvest. It probably also explains the blue garlic halves in your questioner's salad dressing taken from the fridge.

Isoalliin is also the major flavour precursor in onions. They smell different from garlic because they lack alliin and have a second enzyme that intercepts the product of the alliinase reaction to form onions' characteristic tear-producing molecules. Onions do not turn blue because this second reaction leaves less thiosulphinate to be converted to coloured compounds. This explains why onions undergo 'pinking' instead.

Meriel G. Jones
School of Biological Sciences
University of Liverpool, UK

Odour water

If I keep a plastic mineral-water bottle topped up with tap water and regularly drink directly from it, the neck smells vile after a couple of weeks. Why is this and why is it always exactly the same smell?

Ann Gilmour
Belfast, UK

Our mouths are home to around 700 types of bacteria. As well as harmful organisms, which can cause tooth decay, gum disease and permanent bad breath, there are 'good' bacteria, which promote oral health by stopping the harmful ones proliferating.

When you drink directly from a bottle, you leave some of your oral bacteria and saliva on its neck. The saliva contains food debris and dead cells on which oral bacteria can thrive. If you don't wash the neck after you have drunk from the bottle, the bacteria left on the plastic will break down nutrients in the debris and release the unpleasant stale smell your correspondent noticed. The smell is always the same because your bacterial flora stays the same.

This is similar to the situation that causes 'morning breath'. During the night, your saliva flow slows and is less effective at washing out food particles and delivering oxygen to the bacterial flora. This stimulates the growth of anaerobic microbes, which are particularly smelly – hence bad breath in the morning.

Bad breath is likely to be more pronounced if you have been breathing through your mouth, as this will dry out the saliva, further cutting the chances of a good wash-out. One reason for drinking is to wash out a dry mouth, making it particularly likely that material left on the bottle's neck contains problem-causing bacteria and debris.

Joanna Jastrzebska
Auckland, New Zealand

Pooling resources

When I open a new jar of marmalade the contents are a nice, semi-solid, homogeneous mass with a smooth surface, however old the jar is. Yet when I make a spoonful-sized hole in the flat surface to remove some marmalade, the next time I open the jar a couple of days later, the hole has started to fill with a syrupy liquid. What is it about breaking the surface of the marmalade that sets this process in motion? It continues until the jar is empty.

Kenneth Crowther
Derby, UK

A proper marmalade contains plenty of pectin, which is fluid while the product is still hot from cooking but forms a gel as it cools. The gel is a sponge of chain-like pectin molecules in a liquid syrup. The sponge neatly fills the jar as you open it and the syrup neatly fills the sponge, simply because the sponge formed from molecules dispersed evenly throughout the syrup. If you were to skim your marmalade from the top instead of digging great, vulgar holes in it, the marmalade would remain intact.

But if you tear gaps into the delicate structure, quarrying it, then the fluid syrup from the higher levels of sponge will seep down into the hollows.

You might feel guilty though when you remember how forgivingly, selflessly, marmalade turns the other cheek, melting obligingly on hot buttered toast. But don't trust its treacherous meekness. Lumps bide their time to topple onto your best shirt, smearing elbow, table and floor. And in hotels it will humiliate you in the eyes of guests, hosts, clients or colleagues. Can't find that report? What is that sticking to the seat of your trousers?

Jon Richfield
Somerset West, South Africa

Pick-me-up

If you drop a piece of food on the floor, it is supposedly safe to eat it as long as you pick it up before 10 seconds have elapsed, because it takes that amount of time before it can be colonised by microbial life. Is there any truth in this whatsoever?

Lorna Milton
BBC Three Counties Radio
Luton, Bedfordshire, UK

Individual microbes are too small to go crumb-hopping. They travel with whatever medium they are living in or on, in this case whatever dust or dirt is on the floor. When you drop food, two things are likely to happen: traces of the food stick to the floor, and traces of the floor (or what's on the floor) stick to the food. So unless the floor is surgically clean, the food will have acquired a new cargo of bugs however quickly you pick it up.

Chris Newton
Nailsworth, Gloucestershire, UK

This is a polite fiction – everyone knows it is an urban myth but plays along. Jillian Clarke is the youngest recipient of the Ig Nobel prize, won in 2004 for her study while still in high school of the 5-second rule. The time chosen for the 'rule' varies, but she traced its origins to at least as far back as Genghis Khan, when it was the 12-hour rule.

Clarke discovered that the quicker food is scooped off the floor, the fewer bacteria are transferred. Even so, while you would have to be unlucky to get ill, 5 seconds is long enough for food to be contaminated with a lethal dose of *E. coli*.

The number of bacteria that reaches the food depends on various factors: the population density of bacteria on the floor, the contact area between food and floor, and the presence of moisture. Not surprisingly, wet food collects more bacteria than dry food.

That's because, at the microscopic level, food leaves a tread mark because neither food nor floor is perfectly flat. This means that the two surfaces cannot mate perfectly, leaving gaps that bacteria cannot cross. However, if either or both the food or the floor is wet, moisture fills these gaps, allowing bacteria to swim to the food, effectively increasing the contact area. When either the food or floor is wet, there is also a risk that dirt on the floor will adhere to the food.

Mike Follows
Willenhall, West Midlands, UK

Germs need a lot more than 10 seconds to 'colonise' anything, but dropped food needs no formal 'colonisation' to be infectious. Although the number of microbes needed to start an infection differs between diseases, the dropped food can immediately pick up more than necessary.

Let's remember what we carry in on the soles of our shoes, and re-evaluate the 10-second idea. Sprinkle castor sugar thinly onto a rigid surface, then drop a hard sweet onto the sugar and catch it as it bounces. A volume of bacteria to match the sugar visibly clinging to the sweet could infect a ward full of patients – all picked up in a contact that lasted a few milliseconds. A single cell of *Shigella* might be enough to establish an infection, whereas cholera needs perhaps half a million – a barely visible speck.

Jon Richfield
Somerset West, South Africa

The questioner's 10-second rule is not related to the time taken for microbial colonisation. Rather, it is a post-hoc justification for an internal trade-off which we all do all the time. This is because the length of time a food item can remain on a surface and still be deemed consumable is positively related to its desirability and negatively related to the perceived level of contamination of the surface on which it was dropped.

When desirability and perceived contamination are plotted on the same graph, with time on the X axis, the point at which the two lines cross – the 'yuck' point – defines the length of time after which an individual will deem a specific food item dropped on a specific surface to no longer be edible.

For any given surface, the greater the desirability of the food item, the longer the time allowed to pass before it is considered too contaminated. Similarly, a longer time span can elapse when the perceived level of contamination of the surface is lower. As a result, a piece of boiled cabbage dropped on a freshly cleaned kitchen floor is usually deemed inedible the moment it hits the floor, while a piece of chocolate may be deemed safe to eat even after several minutes on grass at a picnic.

For any given food item and surface combination, the desirability and perceived contamination varies between people. It also varies across time for each individual, depending on internal motivation and external pressures. This means that these relationships have varying slopes, and therefore different yuck points, for different people under the same starting conditions.

Thus, for one person, 10 seconds on a living-room carpet may result in an errant chocolate button being binned, while for another, discovering one several days after it rolled under a table may result in a cry of delight followed by a quick brush off before it is popped in the mouth.

Colin Macleod
Glasgow, UK

Let's face it, the 10-second rule is utterly redundant if you drop your sandwich on a pile of dog poo or your chocolate on a recently sterilised kitchen surface.

Bryan McKeen
Dublin, Ireland

Barbecue decay

Over our Christmas barbecue – it was 35 °C – we started an argument. If it was raining (which it wasn't), which would rust faster, a scalding hot barbecue or a cold one, presuming they were made out of an iron-containing material?

Jayne Millington
Perth, Western Australia

If 'rust' is taken to mean the oxides of iron commonly seen on exposed iron and steel surfaces, rather than the oxides which form at several hundred degrees Celsius (unlikely to be attained in a barbecue), then water must be present for rust to develop. A scalding hot barbecue will thus rust far less than a cool one, as rain landing on the former will evaporate instantly.

For the hot barbecue to be moist, the air would have to be saturated at the temperature of the barbecue – a practical impossibility unless the atmospheric pressure could be increased as well.

I really hope no one will try this at home.

Neil Fairweather
Risley, Cheshire, UK

My partner, a blacksmith, says that a barbecue that is scalding hot and remains that way will not rust. If it gets wet while hot and then cools down, however, it will rust faster than a barbecue that gets wet when cold.

Barbecues are generally made of steel, which consists of iron plus a small amount of carbon. If the barbecue is heated to dark red heat – around 600 °C, a temperature easily reached with charcoal – and cools down, some of the carbon burns off. The pure iron left behind rusts easily; rusting can be very quick when the iron is cooled by rain.

A cold barbecue, however, rusts when water gets into cracks in the steel and corrodes the iron, leaving the carbon behind. This is a slow process.

Sarah Lewis-Morgan
Tüngeda, Germany

? Sizzle addiction

I've recently been trying to lose weight and am rather pleased with the results. However, there is one instance every day when the craving for food becomes almost agonising. I have to pass a small food stall in the morning which serves bacon sandwiches. The smell drives me crazy and I'm desperate to buy one, so much so that I've changed my route to work to avoid it. A vegetarian friend also tells me that the one smell that could almost make her start eating meat again is that of bacon grilling. So what has cooking bacon got in it that makes it so tempting?

Peter Hodge
Leicester, UK

I am familiar with these aromas and their effects on the senses. I was once a product development manager in a small ham and bacon processing company in Victoria, Australia.

We frequently offered freshly cooked samples of our products to customers in many of the retail outlets we supplied. Demonstrators were instructed to fry small pieces of bacon, replacing them once they began to look overcooked. This ensured the delectable aroma of freshly cooked bacon was always emanating from the pans.

But why is the smell so good? Cured solid meat products, such as leg and shoulder hams, sides of bacon and beef silversides, to mention just a few, are saturated with a 'curing brine'. This is a solution of salt, nitrite, phosphates, hydrolysed corn starches and sundry flavouring ingredients.

Many saccharides present in hydrolysed corn starches are reducing sugars, which, at the high temperatures of a frying pan or grill, combine with some of the amino acids in the meat in what is known as the Maillard reaction. This is analogous to the caramelisation of sweetened condensed milk when it is heated for long enough.

The products formed in the early stages of these Maillard reactions frequently have pleasant aromas and tastes. As the reactions continue, however, the aromas and tastes of the compounds they produce begin to decline and become quite unpleasant. Demonstrators were instructed to replace well-cooked bacon with fresh to avoid this.

A visual indication of this is when the attractive golden-brown colour gives way to darker colours. These Maillard-derived colours, flavours and aromas are not limited to bacon, although those derived from pork products seem to be much more attractive than those from other meats. For example, ham steaks release very much the same attractive flavours and aromas when they are cooked, but because slices of bacon are thinner and therefore heat through more rapidly, they develop and release their aromas faster than do ham steaks. Thin slices of cured pork sausages yield similarly attractive flavours and aromas when they are cooked.

Dan Smith
Traralgon, Victoria, Australia

I have never knowingly eaten bacon, and when I've smelled it cooking I haven't felt any craving. Likewise, friends who have never eaten it tell me they don't find the smell particularly enticing. So I would suggest it is not something intrinsic in the smell of cooking bacon that makes it irresistible but, rather, that the smell evokes memories of having eaten it. I assume it must taste delicious.

Yonatan Silver
Jerusalem, Israel

Shrink wrapped

In our office, where we test barcode scanners, we have a sample plastic bottle of Coca-Cola that has been left unopened for four years. Over this time the bottle has collapsed to the point where its famous shape is now barely recognisable. Because it is airtight, you might have expected the opposite from a bottle containing a fizzy drink. What could be causing the reduction in size? Might it be something to do with Coke's famously secret ingredients?

Ayrton Nabokov
Melbourne, Australia

Plastic bottles are indeed pressurised. Each 2-litre bottle of Coca-Cola contains approximately 8.6 litres of carbon dioxide when it is manufactured. Plastic, in this case polyethylene terephthalate (PET), is not a perfect barrier to gas and therefore, over a long period, the carbon dioxide will escape through the walls of the bottle.

As the gas escapes, the pressure inside the bottle gradually falls to atmospheric pressure. The volume of the liquid also falls as the dissolved gas escapes from it, which eventually creates a slight vacuum inside the sealed bottle which causes the bottle to distort in shape. This is known in the soft-drink industry as the 'panelling effect'.

Coca-Cola Great Britain press office
London, UK

Kava on the brain

I recently drank some of the interesting root-based drink kava on the Pacific island of Vanuatu. I'm happy to report that it had some odd effects. What exactly did it do to my brain?

Robert Steers
Galston, New South Wales, Australia

Because your correspondent drank the kava in Vanuatu, it may have been made using either fresh or dried roots, while in other countries where kava is drunk – namely, Fiji, Tonga and Samoa – it is almost exclusively prepared from dried roots. Vanuatu kava, especially that from fresh material, is much more potent and its effects tend to be greater.

The active chemical ingredients in kava, called kavalactones, produce a number of effects in both the brain and the rest of the body. Initially the tongue becomes numb, as does the inner lining of the mouth. Some of the other effects may depend on the familiarity of the user with the drink. A novice user may find the drink bitter or sour and that food loses its taste and flavour. Nausea may follow, along with headache and intestinal discomfort, effects not experienced by the habitual drinker.

In contrast to alcohol, kava used in moderate amounts produces a calming effect, reduces fatigue, allays anxiety and stress, and induces a generally pleasant, cheerful and sociable attitude. It is partly for these reasons that it has been consumed in South Pacific communities for hundreds of years as a social drink. One hears expressions like, 'you cannot hate with kava in you' and 'unlike liquor, kava does not provoke aggressive, boisterous or violent behaviour'. Nor does it cause the hangovers, physical addiction, memory loss or diminished reasoning associated with alcohol.

Kavalactones have been shown to produce a number of biological effects in the brain that could account for the above observations. They include the compounds' ability to produce a local anaesthetic-like effect – hence the numbing of the tongue – and to act on drug receptors in a similar way to some anxiety and stress medications such as benzodiazepine. As a result, kava was introduced in the western world to treat anxiety, stress, restlessness and sleep disorders.

However, less positive effects have been reported with the use of excessive amounts of kava, and in a few cases where it has been combined with medical drugs.

There have been some reports of kava causing liver damage in people who live in western nations where it has been used in the form of pills and other such preparations. Consequently, some countries have suspended the sale of kava or issued health advisories. As this medical condition has not been reported in traditional kava drinkers, it is unclear whether it is directly associated with kava itself, or with the manufacturing process or some other factors. In any case, kava is still widely drunk by people from the South Pacific, including myself – I am originally from Fiji.

Yadhu Singh
Brookings, South Dakota, US

Exactly how kava acts on the brain is unknown. Benzodiazepine anti-anxiety drugs work by stimulating the brain's gamma-aminobutyric acid (GABA) receptor proteins, which regulate signal transmission between nerve cells, but kava is not thought to work on GABA. It is also thought to act in a different way to the opioid drugs. It may work on the brain's limbic system, located in the medial temporal lobe of the brain, which is involved in controlling emotions. There has been speculation that it may antagonise the neurotransmitter chemical dopamine.

Jamie Horder
Oxford, UK

Bigger bag

Why, when you pour boiling water on a tea bag, does the bag inflate?

William Hughes-Games
Waipara, New Zealand

The first cause of the tea bag's inflation is likely to be the expansion of air inside the bag as it is heated from room temperature (around 25 °C, or 298 kelvin) to the boiling point of water (100 °C, or 373 kelvin).

According to Charles's law, the volume of a gas is proportional to its absolute (or kelvin) temperature. So the air in the tea bag will expand by a factor of 373 divided by 298, or roughly 1.25, a 25 per cent increase in volume.

A second reason for the inflation could be the phenomenon known as nucleate boiling. At atmospheric pressure water boils at 100 °C, but it is actually quite difficult for bubbles of water vapour, or steam, to form in the bulk of a liquid. Boiling usually occurs only at a solid surface where small cracks and crevices facilitate the formation of bubble nuclei, which then detach and grow as they rise. This is why in a pot of boiling water you will usually see bubbles appearing only at the base or walls.

What this means is that although the bulk of the water is superheated and ready to boil, it is unable to do so until it comes into contact with a rough surface. The leaves in the tea bag would provide ideal nucleating sites for bubbles of steam to form, which would also help to inflate the tea bag.

A third possible source of gas inside the tea bag would be from volatile compounds released from the tea leaves when they are heated, though I suspect this wouldn't be very significant compared to the first two effects.

Simon Iveson
Mayfield East, New South Wales, Australia

The pressure of the inflated air in the tea bag is insufficient to overcome the surface tension of the water held in the minute holes in the tea bag's fabric. So when the air inside expands, it cannot leak out, and inflates the bag instead.

R. Watkins
Marske-by-the-Sea, North Yorkshire, UK

Scraping the bottom

Why does the bottle of red wine vinegar I buy from my supermarket start off without sediment and then acquire it after a couple of months?

Niall O'Sullivan
Liverpool, UK

This could be due to a combination of things. Wines contain tartrates formed from the tartaric acid naturally present in grapes. These are the source of the glass-like crystals which often settle at the bottom of a bottle of wine.

Tartrates become less soluble as the temperature drops, so

these crystals form in wine stored at a lower temperature after bottling than that in the vats used to age the wine. The same should be true of wine vinegar because it is only the alcohol which is soured. The constituents responsible for the flavour of the wine, including tartrates, remain in the aqueous part of the vinegar or wine throughout the fermentation and ageing process.

It may also be the case that some yeasts and bacteria used to ferment the alcohol remain in suspension only to settle out when the wine vinegar is left standing.

Terence Hollingworth
Blagnac, France

As a chemist, I know that red wine contains polyphenolic compounds that are only slightly soluble and, given time, these slowly change composition to give insoluble compounds. This process can happen in the presence or absence of oxygen but is faster with it.

The polyphenolics are oxygenated molecules anyway, which makes them heavier than water, hence the sediment.

Colin Cook
Basildon, Essex, UK

? Part-time curdling

On 3 July last year I opened a 2-litre plastic bottle of semi-skimmed, homogenised milk, with a use-by date of 5 July. I poured out a small cup of what turned out to be curds: a sniff confirmed the milk was off. I guessed that it had been inadequately refrigerated but because I was making pancakes I continued to pour it out, suspecting the sour taste would not be noticed. To my surprise, the remainder was fine and still was a day later: the

curdling was restricted to the top 4 centimetres of the milk, where the bottle was fairly narrow. How can this have happened?

Bob Ladd
Edinburgh, UK

Pasteurisation and, to a degree, homogenisation which evens out the populations of bacteria, reduce the number of bacteria in milk sufficiently to prevent souring for impressive periods, if it is kept cool. Later, it takes very few bacteria to start spoilage, which makes it critically important to keep the equipment sterile between treatment and bottling. That is where most slip-ups occur, not in the pasteurisation itself.

In your case the souring could have resulted from a microscopic amount of, say, *Lactobacillus* or *Streptococcus* inside the cap or neck of the bottle.

Possibly the bottle had been allowed to warm up a little in the shop or when it was being transported. If the bottle was not severely shaken or tipped at any point, it would take a long time for the culture to spread through the bottle, because the species mentioned above do not swim and curdling solidifies the liquid, inhibiting convection currents in the milk.

Also, bacteria consume some of the dissolved solids, such as milk sugar, so the whey would be less dense than the milk and float above it, helping to keep the culture up in the neck of the bottle.

Jon Richfield
Somerset West, South Africa

Bug wash

I want to know if washing vegetables that you would normally eat raw is any protection against illnesses such as the E. coli outbreak in Germany in 2011, thought to have been caused by contaminated salads. It seems that holding the produce under running water from a tap will have limited success in removing contaminants. Am I right?

Peter Mariani
Newcastle, New South Wales, Australia

Any good rinse will reduce surface contamination to a degree. This is sufficient to protect against some of the most serious diseases, such as cholera, which require something like 500,000 bacterial cells to make a moderately susceptible person ill.

Besides, even if enough micro-organisms remain after washing to make you slightly sick, that minor infection could improve the body's chances of building up an effective resistance to that particular pathogen. Then again, if you suspect that apparently clean food might be slightly contaminated, it might help to rinse it with hypochlorite or peroxide before a final rinse in clean water.

Nothing is foolproof though – other microbes, *Shigella* for instance, need only a few cells to establish a dangerous infection, perhaps even just one.

Anyone experienced in handling microbial cultures will know that you cannot wash every germ away with a domestic rinse, so if you have reason to fear a really dangerous infection, rinsing vegetables and salad is certainly inadequate.

For the duration of any outbreak observe the protocol: 'Peel it, cook it, or forget it!' And I am not even sure about the peeling.

Jon Richfield
Somerset West, South Africa

Rinsing fruit and vegetables might have some beneficial effect if the water is chlorinated, but I suspect you would waste a lot of drinking water in the process.

If you are concerned for your vegetables' purity you may do well to wash them in what I believe was called 'pinky paani', as the British did in India during colonial times. This pink water is a dilute solution of Condy's crystals or potassium permanganate. It is an effective disinfectant and seems to have no harmful effect on people, and the extra manganese might be good for one's ligaments.

Robert Maxwell
Brisbane, Australia

Get crackling

If you want pork skin to be crackly when you roast it, recipes suggest rubbing salt into the skin first. What does the salt do?

Peder Olafsson
Bruges, Belgium

This question was particularly pertinent to me, because I read it just before placing a pork joint in the oven.

Very simply, the salt draws water out of the skin cells. This reduces the water content of the skin so less water has to be driven off during cooking. The skin, therefore, cooks relatively quickly when compared with the muscle (the meat) and so it is crisp by the time the joint is cooked.

However, as any chef will tell you, as well as rubbing with salt, the foolproof way to get good crackling is to stick the joint under a hot grill for about 20 minutes before serving.

Gillian Coates
Anglesey, UK

Rubbing salt into pork skin reduces the water content that would otherwise keep the skin soft while cooking. Scoring the skin deeply before applying salt helps, too. Be aware, though, that the high salt content makes this delicacy extremely unhealthy, however tasty. When my dog Sparky got hold of some he promptly, and probably very wisely, threw up.

A healthier alternative is to roast the joint without salting.

Tony Holkham
Boncath, Pembrokeshire, UK

Frosty sludge

If I freeze cakes with icing on them, when I defrost them the icing becomes soft and sticky. Why?

Paula Paine
Plymouth, Devon, UK

The question does not specify whether the icing is fondant, royal or buttercream. However, the general rule of thumb is that all iced cakes should be defrosted uncovered, open to the air at room temperature and not in a fridge (buttercream excepted).

Also, fondant and royal-iced cakes should not be stored in a fridge but instead, loosely covered and kept cool. If these icings, which are also sold 'ready to roll', are stored in a fridge or defrosted while still freezer-wrapped, the icing becomes very sticky – especially fondant icing – due to any excess moisture being absorbed by the high sugar content.

Denise R. Harris-Edwards
Trefor, Anglesey, UK

2 Our bodies

? Blind to the truth?

My mum says that if I read under the bedclothes using a torch I will damage my eyesight. I told her it's an urban myth, but am I right?

Ben Renton (aged 11)
Leicester, UK

Dim light makes us strain our eyes more than in normal lighting because our eye muscles receive mixed signals. It encourages the eye to relax and let in as much light as possible through a large pupil, but also encourages it to contract the pupil to maintain a focused image.

On top of that, the dimmer the light, the lower the contrast between letters and the paper on which they are written, hence our eyes have to work harder to distinguish words on the page. All that results in our eyes becoming tired more quickly than in normal lighting conditions, leading to itchy, dry eyes – through reduced blinking as your eyes try to focus – blurred vision and headaches.

You may also get back and neck aches because under the bedclothes with a torch in your hand is not really the most comfortable position for reading. However, all of these symptoms eventually go away and none of them can damage your eyesight permanently.

Joanna Jastrzebska
Auckland, New Zealand

When I was 11, I often used to read under the covers with a torch. My mum also told me it would damage my eyesight. I'm now 53 and, apart from the inevitable onset of presbyopia – the progressive loss of elasticity in the lenses of the eye that make it increasingly hard to focus on near objects – my eyesight remains very good. However, by the time I was 13 I had persuaded my mum to be more relaxed about the time at which I went to sleep, so I could read with the light on.

Andrew Fogg
Sandy, Bedfordshire, UK

Snail nail

Why do my fingernails grow at least three times as fast as my toenails?

Tony Holkham
Petersfield, Hampshire, UK

Toenails are subject to less wear and tear and so do not need to grow as quickly. Hands can be used as spades, for example, and fingernails can be used to prise things open.

According to Linden Edwards and Ralph Schott, who published a paper on the subject in 1937 in the *Ohio Journal of Science* (vol. 37, p. 91), toenails grow at half the rate of fingernails. On average, fingernails grow a little less than 4 centimetres a year. There is quite a big variation between individuals, depending on heredity, gender, age and how much they exercise. Nails grow faster in the summer.

Mike Follows
Willenhall, West Midlands, UK

Wet weather problem

Why do I find I want to urinate more frequently and more copiously in very cold weather?

Jill Cherry
Mönchengladbach, Germany

It's for the same reason that you feel thirsty after going for a swim. When you go out in the cold, or get into cold water, your body restricts the flow of blood to your peripheries and the skin by constricting peripheral blood vessels. This allows it to preserve heat and maintain core temperature.

One consequence of this is that the total volume of your blood vessels is reduced, which initially means your blood pressure increases. In response, your kidneys excrete more fluid and thus reduce blood volume, restoring your blood pressure to its previous level.

When you come into the warm again, the mechanism works in reverse and you have to either drink more or more water has to be taken up from your gut to compensate. In warm conditions we transpire to keep cool, losing water through our skin and mucous membranes, and in consequence excrete less urine.

Terence Hollingworth
Blagnac, France

On a related theme, we have an answer to the second part of the following question but not the first. Why should running water have this effect? – Ed.

Audio stream

Why does the sound of running water increase the urge to urinate? Is this unique to humans?

Eric Jarvis
Milton Keynes, UK

When I was looking after horses and I wished to encourage a horse to urinate, I either hissed through my teeth or moved the straw bedding around to mimic the sound of water, and it nearly always worked.

Also, in a group of horses if one of them urinates, then several others will do the same, usually to the amusement of their riders.

Marion Bleney
Worcester, UK

Waxed out

What is earwax for?

John Liebmann
London, UK

Earwax, also called cerumen, acts as a cleaning agent for the ear with lubricating and antibacterial properties. Cleaning occurs because the epithelium – the surface layer of skin inside the ear – migrates from eardrum to exit, acting as a conveyor belt carrying dust or dirt out of the ear.

At first, this migration is as slow as fingernail growth but, aided by jaw movement, it accelerates once it reaches the entrance of the ear canal. When it gets to the final third of the ear canal, where cerumen is produced, the conveyor

carries both wax and whatever gunge it has accumulated towards the exit. Cerumen consists of a mixture of watery secretions from sweat glands and more viscous secretions from sebaceous glands. Some 60 per cent is keratin but it also contains dead skin cells, fatty acids, alcohol and cholesterol.

There is more to earwax than meets the eye. It helps anthropologists track patterns of human migration because your genes determine whether you have 'wet' or 'dry' earwax. The absence of water in dry earwax, making it grey and flaky rather than the more common moist golden variety, eliminates evaporative cooling, an advantage to those who evolved in cold climates. The gene that reduces sweat production in these people is also responsible for dry earwax.

Surprisingly, problems with earwax can kill. Scuba divers sometimes have difficulty equalising the pressure in the inner ear to the ambient pressure of the water, which increases as the diver descends. It is usually caused by blocked or narrow Eustachian tubes, but earwax can also be responsible.

In such circumstances, one ear will usually 'clear' before the other. A difference in pressure in each ear induces vertigo, an alarming spinning sensation. I experienced this when diving in a murky quarry. Without any visual cues, it took willpower to suppress my anxiety. A couple of days later a plug of earwax fell out of my ear.

Sports researcher William Morgan has suggested that more than half of scuba divers have experienced potentially life-threatening panic attacks. I suspect ears clogged with earwax might account for some of them.

Mike Follows
Willenhall, West Midlands, UK

Revision tactics

Is it better to stay up late on the night before an exam, learning those last-minute facts, or to get up early and revise in the morning before the exam?

Andrew Maeer
Exmouth, Devon, UK

For 40 years, I have told my students that the day before their exam they should pack some sandwiches and a can of drink and go for a walk in a high and windy place with a friend. Forget all the exam tension, ensure that they are fit, and give all the knowledge they have acquired time to sort itself out.

Yet when I come in on an exam morning at about 7am to check on everything, I invariably find all the students sprawled over the stairs, red-eyed and yawning, in a cloud of (illegal) cigarette smoke, poring over their notebooks. Some of them have been there almost all night. I can't think of a better way to reduce your efficiency, and indeed some fall asleep during the exam.

If you must revise at the last minute, the evening before would be better so that you will be fresh in the morning.

John Anderson
Warsaw, Poland

Neither. It's too late by then. If you don't know it by the night before the exam, there's nothing you can do apart from try to relax a bit and rely on all the hard work you've done before. (You have done all the work before, I hope.)

Gail Volans
Brean, Somerset, UK

Ear whacks

Why does having something pushed into my ear make me cough?

Oliver Pilkington
Ilkley, West Yorkshire, UK

This phenomenon is called Arnold's ear-cough reflex. It occurs in about 2 per cent of the population and was first described in 1832 by Friedrich Arnold, professor of anatomy at Heidelberg University in Germany.

The vagus (Latin for 'wanderer') nerve arises in the brain stem and provides a nerve supply to the external ear canal, larynx, heart, stomach and intestine. Stimulation of the auricular branch of the nerve by objects inserted into the ear canal causes a reflex stimulation of the laryngeal branches of the vagus, which produces the cough in susceptible people.

A variant of Arnold's reflex is vomiting caused by reflex stimulation of the vagal branches supplying the stomach. Wealthy Anglo-Saxons who enjoyed feasting – such as the ealdormen who ran local affairs for the king – are said to have poured cold water into the ear to induce vomiting when they had eaten their fill so that they could continue to indulge themselves, hence an alternative name for the auricular nerve: the alderman's nerve. The Romans are said to have achieved the same result during their orgies of eating by tickling the ear canal with a feather.

Maurice Little
Maidstone, Kent, UK

This is one example of what are called neural reflexes. They are generally explained as the 'confusion' of one nerve path (usually sensory) with another (usually motor). For example, rubbing the skin at the back of the neck produces a widening of the pupils, and scratching the skin of a man's inner thigh will result in the raising of the testicle within the scrotum on that side. This kind of information is used clinically to determine the integrity of a particular neural pathway. It can also be used to liven up otherwise dull parties.

Joseph F. Gennaro
University of Florida Medical Center
Gainesville, Florida, US

When scuba diving in the tropics a couple of years ago I had the sensation that something had entered my ear at high speed. I don't recall coughing but I was so shocked I spat out my mouthpiece.

I hastily replaced it and poked around in my ear to remove what had darted in there. When I saw the characteristic iridescent blue stripes of the bluestreak cleaner wrasse (*Labroides dimidiatus*), which cleans bigger fish for food, I assumed that I had just passed over one of its cleaning stations and been relieved of some earwax.

Mike Follows
Willenhall, West Midlands, UK

Eyes down

How do bifocal contact lenses work? Specifically how is the reading part of the lens held in the correct position?

Gino Trulli
Terracina, Italy

There are at least three kinds of bifocal contact lens: translating, concentric and aspheric. The question relates mainly to the translating type, because these are the only ones that have to be positioned correctly all the time. A translating bifocal lens is in two halves, with the part that corrects for short-sightedness at the top and that for long-sightedness at the bottom, pretty much the same design as for bifocal glasses.

Such lenses are relatively small, covering only the centre part of the eye, and they are made of a more rigid material than the more usual soft lenses.

In addition, a small segment of the lens is cut away at the bottom leaving a flat edge. The flat edge ensures that the lens stays correctly aligned when you blink. Its small size and rigidity allow the eyelid to hold the lens centrally while the eyeball translates – slithering up and down between far and near vision.

In concentric lenses, the centre is made up of concentric bands of material that alternate between correction for short sight and long sight. This design exploits the adaptability of the human brain, which can keep a mixture of both near and distant objects in focus at the same time.

The wearers 'train' their brains to 'see' either close-up objects or distant ones as required, because in effect the eye is registering both at the same time. Such lenses move with the eyeball and it doesn't matter if they rotate on the surface of the eye because they are symmetrical.

Finally, aspheric lenses are similar to the progressive lenses in glasses: correction for short sight is in the centre of the lens, and it progressively changes to correction for long sight towards the outside. Once again the brain has to sort out conflicting information from the two parts of the lens.

People who wear progressive lenses will remember the first time they tried them, when head movements produced a wildly oscillating, completely disorientating visual field. Somehow the brain sorts this out and, after a few hours, everything returns to normal and perfect vision ensues.

Terence Hollingworth
Blagnac, France

I wear toric contact lenses, which correct for my astigmatism as well as my short-sightedness. Because these lenses are not symmetrical they have a heavier portion at the bottom designed to 'drag' the lens into the correct position after insertion. The lens itself also has a small mark at the six o'clock position so I can insert it the correct way up when I start wearing them. This mark is invisible to the wearer once the lens is on the eye.

Rhian Chapman
Luton, Bedfordshire, UK

A balanced diet

Are all calories (from all food types) equal in terms of causing weight gain? That is, if I eat complex carbohydrates to the value of 1000 calories, are these as likely to increase my body fat as eating fat or sugar to the value of 1000 calories?

By email, no name or address supplied

Yes and no. The calories are the same. Once your body has converted the consumable compounds to a common compound (acetate, say) then there is no difference between the foods that were converted, whether protein, fat or carbohydrate.

However, not all food compounds are equally well converted, absorbed and assimilated. Many starches are poorly digestible, such as polysaccharides. A great deal of unrefined food passes right through your gut in sheaths of collagen, cellulose and similar poorly digestible natural plastics. Even animals that can digest these with the help of microbes usually need to digest the food twice, either by regurgitating the food after the first pass into the stomach, or

by eating it again once it has been excreted. Ask any rabbit or termite about coprophagy.

Jon Richfield
Somerset West, South Africa

If you restrict the number of calories in your diet, you will not show any difference in weight loss whether you restrict calories from fat or calories from carbohydrates. This has been studied fairly extensively. An example study is 'Comparison of Weight-Loss Diets with Different Compositions of Fat, Protein, and Carbohydrates' (*The New England Journal of Medicine*, vol. 360, p. 859). The researchers found there was no significant difference in weight loss or in waist circumference for people eating diets that contained differing proportions of carbohydrates, fat and protein.

However, this does not mean you will experience the same success with a diet that restricts fat as with one that restricts carbohydrates. We react differently to different foods, and fats help us feel satiated. In this study, subjects only ate what the researchers gave them, but you will not have this type of enforcement.

A balanced, calorie-restricted diet is the best way to diet. Too little fat and you will often feel hungry; too little carbohydrate and you will feel tired and weak. Also, keep in mind that high-fat, high-protein diets place you at a greater risk of heart disease.

Araneae
By email, no address supplied

The little people

Sometimes when I am concentrating on someone who is talking to me, they seem to become very small and appear as a miniature of their normal selves. Everything else in my field of view is also miniaturised and I can control the effect, which is great fun. This is more likely to happen if I'm tired. What is happening and how common is this effect?

Steve Brown
Addlestone, Surrey, UK

Although there is no definitive answer to the question here, the editor has a vested interest in asking it because he also experienced this as a child, usually when reading a book late at night while being very tired. The book pages and text were suddenly far away and very tiny but still each word remained clear. It all sounds rather extraordinary, but if anybody knows the definitive answer, do get in touch – maybe we can set up an international support group for those who have lost the ability, as many people seem to do in adulthood – Ed.

I have experienced this phenomenon for as long as I can remember, but more when I was a child. I have often tried to explain it to people, even as it was happening, but no one I've met has experienced the same thing. How wonderful that this happens to someone else.

Helen Ferrier
Norwich, Norfolk, UK

I experienced the same phenomenon as a young child. I can vividly recall looking down into Trafalgar Square and seeing what I perceived to be miniature people scurrying around. I would have been around three years old and I pointed them out to my uncle and asked him if he could see them too. I was told not to be silly and that the people we could see were all of normal size.

As I remember, I put the reason that he couldn't see them down to the fact that he was a grown-up. It was obvious to me that only children could see these little people. I didn't experience the phenomenon very often, but I have a very poignant memory of catching a glimpse of a little person disappearing into a station. He seemed in an instant to jump from being a little person to normal size and I knew as I saw him that I wouldn't see any more little people, because I was growing up. I have often thought of the experience, but I have never spoken or written about it.

Your question confirms my very unscientific conviction that I did see little people as a child and only wish I could regain that magical outlook on the world.

Ian Buchanan
Hook Norton, Oxfordshire, UK

The two letters above were selected from many others. It is clear that the effect is quite common in childhood, even though most correspondents never knew that other people shared their experience. We can find no scientific studies of the phenomenon, but suspect that the letter below may provide an explanation – Ed.

I have been familiar with this since I was a child, when it used to frighten me. In bed at night the pictures on the wall would dwindle to postage stamp-size. This happened suddenly – they did not zoom out gradually. It happened without any loss of detail or resolution. As an adult I can make it happen by concentrating in a certain way and this can enliven dull meetings at work.

The mechanism may derive from the way that the brain corrects sizes. This is the visual system that adjusts the perceived size of objects to the size you know they should be. To demonstrate this, hold your left hand level with your shoulder about 15 centimetres from your face, and fully extend your right arm. Transfer your gaze from one hand to the other. Both will appear about the same size, although the image of one on the retina will be three or four times as large as the other. It is easy to show in the lab that size correction is done by expanding the smaller image, and this accounts for several well-known optical illusions.

Presumably the mechanism is mediated by particular cells in the visual cortex, and if these tire or are inhibited by higher centres, everything in the visual field will look small.

Galen Ives
Sheffield, South Yorkshire, UK

The well-known Moon illusion provides another example of a distance scaling effect. When the Moon is high in the sky and there are no distance cues it appears small. When it sinks to the horizon and is seen through buildings, trees, and so on, the brain scales it to these closer objects and it appears much larger.

Similarly, when walking in the mountains in thick mist, which removes distance cues, a towering mountain seen through the mist can turn into a nearby rock a few paces later. The brain's ability to adjust our sensation of size from the real size of the image on our retina also explains why so many amateur photographs are disappointing. The people we felt were nearby turn out to be mere dots on our photograph.

Had the earlier correspondent photographed his view of Trafalgar Square, it might well have looked much more like his childish view of it than his uncle's adult perception. The figures' projection on the retina would really have been tiny; it is our brains' interpretation that makes them seem larger. Any of the many books by Richard Gregory on illusions will provide further reading – Ed.

Dim lit

I was reading in very dim light last night and found that I could not read black text on a white background (the character size was about 8 point) but could read white on a black background. I'm 48 and wear reading glasses, which I did not have with me at the time. I expected not to be able to read black-on-white text, but was surprised I could read the opposite. Why is this? It seems counterintuitive.

Ben Deighton
Halifax, West Yorkshire, UK

There are two phenomena which conflate to allow this to happen, by altering the perceived text size and word shape. Your questioner has presbyopia, an age-related diminished ability to focus on near objects created by a lack of elasticity in the lenses of the eyes. Without reading glasses, images are blurred. This is exacerbated in low light as the pupil enlarges to let more light into the eye and thus shortens the depth of field of clear vision. This is familiar to photographers, who alter the depth of field by changing a camera's aperture size to blur the background or make it sharp; the smaller the aperture, the clearer the background. Lenses focus light, not the lack of it. So as your questioner's eyes focus light poorly, the white light from the background will impinge on the edges of black text, making it appear smaller and more blurred. Conversely, when viewing white text on a black background the diffuse light from the text impinges on the black background, producing blurred but perceptually larger text. This makes it easier to read than the black text.

The shape of the 'larger' white words is also more discernible than the shape of the smaller black text. The importance of the overall shape of words to how a skilled reader perceives them has long been debated by cognitive psychologists.

An analogy is to imagine the difficulty of perceiving black specks in a sunlit sky. Compare that with perceiving white specks of the same size on a black background, for example, stars on a clear, moonless night. Even without glasses, a myopic (or short-sighted) person like me can see the diffuse light of the stars as blurry blotches, and can recognise constellations because I am already familiar with their shape, just as your correspondent will be familiar with the shapes of the words on the page.

The latest cellphone models frequently use white text on a dark background. I assume that this is a means of saving battery charge, rather than any specific benefit to the hyperopic (long-sighted) or presbyopic, and is nothing more than serendipitous.

David Muir
Science department
Portobello High School
Edinburgh, UK

Think again

My 14-year-old daughter asked me what language people who are deaf from birth think in. Does anybody know how this cognitive feature develops in deaf people?

Mike Dunn
Sherborne, Dorset, UK

It varies. Deaf and hearing children raised by deaf parents who use sign language will acquire that language in a similar way to the acquisition of speech by hearing children. They will use this language to think with just as any other child will.

However, most deaf children are born to hearing parents who cannot sign, and they will not develop language in the same way. These children often use gestures for basic communication but this is not the same as sign language, and their first exposure to an accessible language may be at school, or later, and this is clearly disadvantageous.

Language is crucial to cognitive development, but what language is used and which modality it is in really doesn't matter.

Rachel Mapson
Edinburgh, UK

The idea that thoughts necessarily come in the form of words seems to have a hard time dying. This misconception appears to be especially common in the English-speaking world, perhaps because most people there are monolingual.

As a polyglot, I can assure you that I don't think in words. I am capable of relating the contents of an article or TV documentary without remembering what language I originally read or saw it in. To me this is proof that the information was not stored in my brain in the form of words.

Even a monolingual person should be aware that relating the contents of an article or discussion is easy, while remembering the exact words might be practically impossible; further evidence that we don't store information as words, but rather as abstract thoughts.

Thoughts only get put into words, or signs, when you want to communicate them or when thinking about communicating them, in which case you might indeed think in words for a while.

Frank Siegrist
Lausanne, Switzerland

Heated debate

High blood pressure can be reduced by taking a hot bath, but why does one's pulse quicken when relaxing in this way? And is high blood pressure related to a faster pulse in any way?

R. Hazelwood
Guildford, Surrey, UK

Your body does not contain enough blood to fill all your blood vessels at the same time, so blood flow is regulated to different parts of the body as needed – to your gut after a meal, to your muscles when exercising, or to your skin when you are hot. Your blood pressure (BP) is calculated by multiplying the volume of blood pumped out by the heart per minute (the cardiac output, or CO) and the total resistance to that flow in your blood vessels (the systemic vascular resistance, or SVR). Thus BP = CO × SVR.

When you step into a hot bath your blood vessels dilate to help you lose heat – you see this as a reddening of the skin – and your SVR falls, hence you experience an immediate fall in blood pressure. This reduction is detected by baroreceptors in your carotid body. These are a type of 'stretch receptor' found in the blood vessels located on each side of your throat, and they control blood pressure by constantly reporting to your brain. Their aim is to keep BP at its 'normal' level by increasing or decreasing CO. This, in turn, is a product of heart rate (HR) and stroke volume (SV), which is the volume of blood that your heart pumps out each beat, typically 55 to 100 millilitres. So CO = HR × SV. You will easily detect a rise in HR by your pulse quickening, and feel an increase in SV as a pounding in your chest, most commonly during exercise.

A faster pulse is not necessarily related to high blood pressure. Combining the above equations shows that heart rate, stroke volume and vascular resistance are all variable factors in determining blood pressure. The 'fight or flight' response, which creates an adrenalin rush in humans, raises all three variables, for example, whereas a hot bath or blood loss will reduce only one.

Longer-term control of blood pressure is mediated largely by the kidneys and I'm afraid to say that once you cool down from your relaxing hot bath your blood pressure will largely return to normal, unless your high blood pressure was stress-induced in the first place.

Roger McMorrow
Consultant anaesthetist
National Maternity & St Vincent's University Hospitals
Dublin, Ireland

Finger roll

Why is it that regardless of whether you are left-handed, right-handed or ambidextrous, you find it much easier to drum your fingers starting with your little finger rather than your forefinger?

Dave Scates
Pyrford, Surrey, UK

Thanks to all of those who offered anatomical explanations, but it seems that they are misplaced. We have uncovered an important subset of the population, some of whom are clearly in need of reassurance – Ed.

It is totally alien to me to attempt to drum my fingers starting with the little finger. Please tell me I am not alone. My husband often says he doesn't know which planet I am from.

Gillian Spencer
Leeds, West Yorkshire, UK

While reading the question I felt the need to drum my fingers and realised I was rolling in the opposite direction to the questioner's theory – starting on my forefinger and finishing at my little one.

For the record, I am right-handed and always drum my fingers with my right hand but I assume that my first finger drum was picked up from the action of another person, rather than being instinctive. To test my theory I asked a few of my friends and I got a good mixture of directions of finger roll. These also indicated that the action came from different directions regardless of whether the performer was left-handed, right-handed or ambidextrous.

Rachel Wickerson
Groombridge, Kent, UK

Two of a kind

Why do we breathe through our mouth or our nose, but not usually both at the same time?

Michael McCullough
London, UK

Like a fluid, if given a choice of paths air will take the widest and straightest.

To breathe, our lungs expand and contract, creating a pressure difference between the windpipe and the outside air. If the mouth is shut, there is only one path, through the nose, and air flows that way. But as soon as we open our mouths, there are two pathways, one wide and straight via the mouth and the other narrow and twisted via the nose.

Some air still flows through the nose, but it is vastly exceeded by that through the mouth, so we seem to be breathing through the mouth only.

Alec Cawley
Newbury, Berkshire, UK

Magpies

Why do we like shiny things?

Jonah Lawton
London, UK

Arguments on this point are necessarily speculative, but it's worth noting that we associate shininess, cleanliness and crisp outlines with objectively favourable attributes. In assessing a mate, a companion or a rival, we spontaneously see bright eyes and teeth, glowing skin and glossy hair as signs of health and quality. As children, we like things that stimulate our nervous systems with clear, vivid colours, contrasts and light.

Art may be seen as a form of play behaviour, in that it relies on elements that matter to our mental and physical development. As adults, our senses and creativity put a premium on media and themes that stimulate our innate mental systems in important ways.

Shiny things present intense, characteristic stimuli, and are used in social signals and communication, even in creatures that do not see art in our terms. Such dramatic signals may be based on anatomy or physiology, such as peacock tails or the belling of a stag, or may be collected and arranged as adornments, like bowerbird displays and human medals or finery. Much as we enjoy speech, we enjoy communication by vivid stimuli in a broader range of contexts.

Jon Richfield
Somerset West, South Africa

We like shiny things for evolutionary reasons. The shiniest thing our primitive ancestors saw was probably sunlit water. Those who were attracted to it must surely have increased their survival chances, described as follows in my poem:

All that glisters
Why should we prize a shining thing,
A sapphire, or a diamond ring?
Diamonds of light and water glint
Through leafy trees. Just here! They hint
For a picnic spot beside a stream
To dream a retrospective dream.
If we stay here we will survive
Or children and descendants thrive.
The rushing stream is crystal clear,
We'll be OK if we stay here.
Sparkles signal, stay alive!
With light and water, all life thrives.
Here food and sheltering trees too grow,
Even if we do not know
That's why we prize a shining thing,
A sapphire, or a diamond ring

Wendy Shutler
By email, no address supplied

Getting lippy

As soon as the temperature falls below about 5 °C I have to avoid smiling because, if I do, my lips crack and bleed. Why does this happen and am I the only person to suffer in this way?

Pascal Rinaldi
Perpignan, France

Cold winter air cannot hold much moisture, so when it is warmed indoors to a comfortable temperature it will be drier than summer air of comparable warmth. This causes the skin to dry out more readily, rendering it less elastic and more prone to chapping.

Skin also expands and contracts with rising and falling temperature. This means the greater difference between indoor and outdoor temperatures in winter increases the stress placed on the skin as you go in and outside. Lips are particularly prone to chapping because they lack the oil glands present on the rest of the skin whose sebum acts as a natural barrier to desiccation and helps to keep it supple. Repeated wetting of the lips through licking only makes things worse by stepping up evaporative cooling.

Lips are also frequently flexed in the course of eating, talking and changing expressions, and, in an already dried state, acidic and salty foods only inflame them more. Smiling provides the final insult because it pulls the corners of the mouth considerably, stretching lips to breaking point.

Why 5 °C should be so critical in your case may require a little speculation, but the most likely explanation is that with the encroaching winter your skin has already dried out somewhat, and that as the biting air chills towards zero the last of any moisture it is holding condenses as frost.

Be reassured that you are far from alone in this, and the problem is relatively easy to remedy. Start applying lip balm or moisturiser regularly before the winter really sets in, and aim to wear a scarf that covers your mouth.

It is the same dry winter air that aggravates sinus headaches and raw coughs. Some remedies are simple – many householders find that humidifiers hooked over their radiators or having pot plants in the home provide relief.

Even an underlying medical condition should be readily treatable. It could just be that your lifestyle exposes you more than other people to the various factors. Perhaps they are the ones who need to get out more.

Len Winokur
Leeds, West Yorkshire, UK

Mole hunt

I have a mole out of which hairs grow much longer than in the surrounding skin. Why do some moles promote longer hair growth?

David Chang
London, UK

There is more to hair growth than meets the eye. An eyelash, for example, is a marvel of programmed curvature, texture, colour, thickness, length and persistence. The hairs on a porcupine are a microscopic and macroscopic tour de force. Even the growth of simple human body hair requires elaborate control of tissues in the papilla at the base of the follicle from which any hair sprouts. Such control demands interaction with surrounding cells and the body's hormones.

Your long mole hair probably grows from papillae that originally produced fine, short vellus hair or humdrum body hair. The mole tissue grew from skin cells, mainly melanocytes, whose own growth controls had been disrupted by genetic or epigenetic changes to their DNA. They are now less inhibited and grow according to a programme differing from those of surrounding tissues, so their cells' textures and physiological products affect the growth of their neighbouring cells in turn. In particular, they can disrupt the activity of papillae, making them produce hairs unchecked. These hairs are often thicker and more simply structured than normal hair.

By contrast, a normal hair stops growing at shorter intervals, pauses for a suitable time, and is then shed after a few months or years of service.

Jon Richfield
Somerset West, South Africa

Write or wrong?

Why do we have different styles of handwriting? It is so individual that you can identify people such as friends and colleagues purely from their handwriting.

Colin Mascord
Sydney, Australia

Handwriting is influenced by factors including the style we were first taught, how our minds reacted to this teaching, how we learned from errors, whether we are left or right-handed, the anatomy of our hands, the nature, quality and dexterity of our hand-eye coordination, the effect of our personalities on the conscious and subconscious choices we make about our handwriting as we grow up, plus various cultural influences we absorb over time.

Given these influences it's more surprising that we are able to use handwriting to communicate at all. I've had to type this because my handwriting's terrible.

Geoff Convery
Gainsborough, Lincolnshire, UK

My friend writes on a blackboard using letters 10 times bigger than on paper, yet I can still recognise his handwriting. So it is not just a matter of small muscle control.

Philip Roe
By email, no address supplied

3 Plants and animals

Up with the lark

Why do birds sing as dawn breaks? Indeed, why do some sing at dusk? And for what reasons do they eventually stop?

Eva Sanz
Tarragona, Spain

The main function of most birdsong is long-distance communication, either to mark territory or to be sociable. As such it is largely intraspecific; blackbirds sing to impress blackbirds, not buntings. In contrast, social vocalisation, such as coordinating group activity, largely occurs at short range during active flight or foraging, or when settling down for the night or preparing to take flight as a flock.

Like any form of communication, birdsong bears an energy cost and requires channel capacity, which is limited largely by background noise and the quality of the medium, in this case air. In the mornings and evenings the air tends to be still, which reduces competing noise. It is also cooler at low altitudes, which favours transmission of clear sounds. Also, few birds forage at dusk, so in terms of energy use that is an economical time to perform.

Because much birdsong is territorial, it is practical for each species to sing at fixed times to avoid wasting energy on talking when no one is listening or when other species are competing for air time. Ideally, that male blackbird would be saying: 'If you are a male, keep off, or else! But if you are a female, let's get together.' Later, when other species are singing, he can go off to catch the early, deaf, worm.

Antony David
London, UK

Birds sing more at dawn and dusk than at other times because that is when they can hear more birds singing. Frequently at these times the wind drops and a temperature inversion forms – this is a layer of warmer air above cooler air. This changes the way in which sound is carried through the air, refracting sound waves back towards the ground that otherwise would have dissipated in the air. The upshot is that sound is carried further at dawn and dusk.

Thus, if a bird devotes most of the energy it spends on singing to those times, it is heard by the widest possible audience. Of course, there are birds that can be heard singing at any time of day, but even these will tend to sing more at dawn and dusk when there is a temperature inversion.

Nigel Depledge
Spennymoor, Durham, UK

? Blacksmith's dilemma

Why did American cowboys need to shoe their horses, but the Native Americans did not?

Larry Curley
Huntingdon, Cambridgeshire, UK

The answer to this stems from evolution; not just biological evolution but also variations in technological and social evolution on both sides of the Atlantic.

The horse's native habitat is large grassy plains with a

generally dry climate, such as the steppes of central Asia, where the wild ass originated, the African savannah, which is home to near-relatives such as the zebra and the now-extinct quagga, and the prairies of North America where the genus *Equus* evolved.

Horses were driven to extinction in North America about 7,600 years ago, possibly by climate change or hunting by the ancestors of Native Americans. They only returned to the New World when the Spanish brought them there in the 16th century.

The tribes of the North American plains and the American Southwest came across these horses, or at least their feral descendants, after they began to escape from Europeans around 1540. Newly established wild herds spread up the Mississippi valley, where most of the tribes had a settled agrarian lifestyle. The Plains Indians led a nomadic existence, and despite never having seen a European – mounted on a horse or otherwise – it was they who realised the horse's potential for enhanced mobility.

First to mount up were the Kiowa and tribes of the Missouri valley, who were riding horses by the 1680s. By 1714 the Comanche of Wyoming had joined them on horseback, followed by the Snake of southern Idaho and eastern Oregon, and the Cheyenne of Minnesota and North Dakota, who in the 1730s introduced horsemanship to their neighbours, the Teton Sioux. Finally, the Sarcee tribe of Canada became the most northerly mounted tribe by 1784.

In the space of just over a century the horse had transformed Native North American society, and not always for the better. The arable society of the Missouri valley was ultimately destroyed by raiding 'war parties' of Comanche, Cheyenne and Dakotas, well before Europeans began their genocidal march.

The Plains Indians were hunter-gatherers, with no

significant metal-working skills, so any metal goods were obtained through trade with Europeans and hence were at a premium. Even if they had needed them, horseshoes would always have been a technology beyond their socio-economic resources. But they had no need to shoe their horses.

The horseshoe was developed to meet the conditions faced by domestic horses in north-west Europe, where, judging from archaeological evidence, they were probably first produced by the Gauls or Franks in the 5th century. Europeans needed horseshoes because of a combination of climate, terrain and pattern of use, with the generally wet weather and soft, heavy soils acting to soften the normally calloused sole of the hoof. Horses were used for travel and in wars. They were often heavily laden while travelling at quite high speeds, which placed great stress on the hooves, often causing them to wear unevenly and eventually split, rendering the animal lame and useless.

The lifestyles of horses used by Plains Indians, on the other hand, differed little from that of their ancestors in the wild. The animals moved together in large numbers at relatively low speeds, over flat, arid steppe country. As a result, their hooves were harder and wore more evenly. In addition, Native American warriors had more than one mount each, with one band of 2,000 Comanche braves keeping a string of 15,000 horses in tow.

The quality of husbandry among European settlers often left a great deal to be desired. The US cavalryman at the time of the Indian wars in the 18th and 19th centuries was usually an indifferent horseman, and more concerned with his own well-being than that of his mount, which was after all government property.

By contrast a Plains Indian brave's horses were his fortune and livelihood, and he cared for them accordingly if he valued his life. Cowboys riding the range were in a very similar position, which was why horse-stealing was considered the worst of all crimes in the old west, as it was tantamount to murder.

However, the peculiarities of their profession, and the specific qualities demanded from the horse required the use of shoes. The classic cowboy's mount was the quarter horse, the fastest steed in the world, but only over short distances (the quarter-mile that gave it its name). This enabled a cowboy to race from one point around a large herd to another at short notice and in short order, but applied stresses a bare hoof could not sustain in the long term.

Hadrian Jeffs
Norwich, Norfolk, UK

Fear of height

If I saw a huge dinosaur, I would probably run for my life. So why do ants seem oblivious to a human towering over them? Do ants not get scared?

Robert Watson
Jesmond, New South Wales, Australia

I think it's all relative. I measured the height of several local ants and found the largest black carpenter ants reach an average of 5 millimetres above the ground. The tallest dinosaurs were about 8 metres high. Humans are mostly less than 2 metres tall. This would make the largest dinosaur about four times the height of an average human while the average human would be about 1,000 times the size of an ant.

Some ants do, in fact, seem to sense us – especially if our shadows fall over them – and run. But mostly I suspect we are just too big to enter their awareness. Also, I wonder if ants ever look up.

Earle McNeil
Olympia, Washington, US

Ants' attitude to life is vastly different to that of mammals, which invest a great deal of time and energy in their young and have evolved numerous means of self-protection. Ants, on the other hand, invest very little in their myriad workers, all being easily replaceable clones, so they have no individual fear of being killed. If the nest is threatened, however, it's a different story, and they will defend it to the death.

Also, ants have been around for more than 100 million years. If they think about it at all, which is unlikely, no doubt ants would see humans as a very transient species, occupying but a moment in time on their planet.

Tony Holkham
Boncath, Pembrokeshire, UK

? Dirty dining

Why do flies like eating dog turds? To me, this seems horrible. Why don't they get sick like a human would?

Cindy Germond
Sydney, Australia

This is a larger issue than dog turds. Parents worldwide teach children faeces are dirty, messy, bad and yucky. Thus the stuff itself and all the synonyms for it that we learn as children become lodged in the brain as a bad thing. We do not play with excrement. This extends to all excrement from any source.

Through this process we learn from a young age that excrement may harbour 'germs', which is sometimes true, although not all microbes are germs and therefore bad for you. Nonetheless, the lesson we learn is a good one because much gastrointestinal disease is transmitted by the so-called 'faecal-oral' route, making hand-washing after visiting the toilet imperative.

Because I was a paediatric gastroenterologist for almost 40 years, I have looked through the microscope at thousands of specimens of human faeces, searching for evidence of maldigestion (undigested fat or muscle), or more commonly inflammation (white and red blood cells). At the microscopic level, excrement is a seething mass of microbes – bacteria, archaea, yeast, and occasionally visible protozoa or fungi all scooting around either actively or by Brownian movement among the debris of digested food particles and cellular cast-offs from the bowel. The microscopic view is almost a work of art as the energetic micro-world recycles debris into life forms that will in turn become food for something further up the food chain.

So to answer the question: flies were never instructed by their parents to avoid faeces, dog or human, and neither were African dung beetles. So they do not object to the odours, hydrogen sulphide, dimethyl sulphide and similar compounds that have been formed by sulphur-metabolising microbes in our colon, or to the appearance of the turd itself. They just see a buffet where much of the digestive work has already been done and they can feed, lay their eggs and relax.

Maybe the Lord of the Flies gets to enjoy the biggest dump.

Adrian Jones
Professor Emeritus, Pediatric Gastroenterology and Nutrition
University of Alberta, Edmonton, Canada

Sadly, in our mindless quest for economic growth, humans rarely appreciate the services the natural world provides free of charge and instead we often seek to trash it.

Luckily for us – or perhaps unluckily when it's our favourite item of clothing – insects are less discerning and will eat any kind of organic matter, whether it be fur, dead vegetation or excrement. For their part insects profit from a ready supply of food and energy, while we profit because such waste products, which otherwise would build up and bury us, are recycled relatively rapidly back into the food chain.

As far as faeces are concerned, despite having passed through a digestive tract, they still contain enough organic material to feed the larvae of flies, and often the adult flies as well. Effectively, dog faeces are composed of material which the dog has not digested, just as there is organic matter in our faeces which humans can't digest – think sweetcorn. Flies are able to digest the remainder.

Insects which live in environments with a high microbial load have evolved a resistance to potentially harmful microbes. For instance, they have genes that govern the synthesis of antimicrobial peptides, which, to them, are effectively broad-spectrum antibiotics.

Dogs and humans play host to a great variety of bacteria in the gut which are actually beneficial. The superbug bacterium *Clostridium difficile* may be present in a healthy gut without causing harm but it can cause problems for hospital patients whose immune systems are weak. A controversial way of restoring gut health after an episode of diarrhoea, say, is to inject a sample of faeces from a close relative into the colon.

So it is not necessarily true that people would fall ill eating dog faeces. I would not recommend the practice, however, because one major danger is that they may contain eggs of flat or roundworms of the genus *Toxocara*, which present no threat to insects but which can establish themselves in humans.

Nonetheless, in cities with many dogs, and particularly in winter when insects feed on faeces less, we can unwittingly ingest a significant amount of airborne bacteria from that source.

Terence Hollingworth
Blagnac, France

Tree healer

The trunk of a small fir tree in my garden broke and the tree fell over. More in hope than expectation I pulled the top part back up, taped the halves back together, and screwed in metal plates to hold them in place. To my amazement the tree survived and, months later, when I removed the tape and the plates, the break had healed and the trunk was whole again. How do trees repair such trauma? Can all tree species do it?

Alan Crowther
Birmingham, UK

Well done. I would have bet against such a successful outcome, especially with a conifer. At such an injury they often produce enough resin to smother the tissues and strangle the tree.

Essentially you connected the cambium, the growing tissue under the bark, at the area of the break. It grew rapidly, joining the edges nicely. This is what one aims for when grafting a bud or stem onto a plant: joining cambium to cambium, keeping it moist, protected, and generally supported until it has grown enough structural tissue to hold it in place and enough conductive tissue to keep the graft fed.

A word of warning. Your tree has grown good supporting tissue around the trunk and, as it grows larger, that layer of new growth will become stronger; in future years the tree should be as good as new. However, the older wood inside the break will never mend.

For now, your tree is being held up by little more than a skin of new wood. Give it flexible support for a few years in case a minor bump or strong wind breaks it again. Rigid support discourages the growth of strong buttress tissue.

Jon Richfield
Somerset West, South Africa

Your questioner has rediscovered the principle of grafting. Not only can most trees do this but, given the right circumstances, other plants can too. Many fruit trees and ornamental trees and shrubs, including roses, are propagated through grafting; even some vegetables, such as tomatoes, can be grafted.

Grafting is often used to combine desirable qualities from the rootstock, such as dwarfing or disease resistance, with desirable attributes from the scion – the branch or stalk being added – which might include fruit quality or desirable flowers. It can be an effective method of propagation. It also occurs naturally: where stems or roots of the same species grow together they will unite.

Two points are essential for successful grafting: the cambial layers of the scion and rootstock must be brought together; and both halves must be kept alive long enough for the union to form. Often the scion is removed from its own roots and, in practice, keeping the scion alive means preventing it from drying out. In the case of the questioner's tree it sounds as if the trunk didn't break completely, so putting the top back in place would satisfy the first requirement and probably enough of the vascular system remained functioning for the natural supply of water to the top to continue.

The process depends on the fact that, generally, mature plant cells in vegetative tissue have a much wider ability to grow and differentiate than do animal cells. So once the grafter has put the two halves in place, new tissues form that bind the parts together. That said, where large grafts are concerned the vascular tissue will often be more effective in regeneration than the internal wood, so a graft can remain a weak point.

Grafting has other uses too. In my career as a plant pathologist I have used several versions to transmit plant pathogens that are difficult or inconvenient to spread in other ways.

Grafting is not always successful, but the reasons may not necessarily be the obvious ones of the scion being the wrong size or drying out. In walnuts, blackline disease, which causes some trees to die several years after apparently successful propagation, is due to a virus in one part causing a necrotic reaction in the other part as it moves across the graft union.

The bottom line for gardeners is that when a stem or branch breaks in a tree or shrub you would rather keep, you don't need to immediately dig it out and rush off to buy a replacement. Do as the questioner did and consider trying to repair it by self-grafting.

If you want to know more about grafting, any edition of *The Grafter's Handbook* by R. J. Garner, first published in 1947, contains a wealth of information.

D. J. Barbara
Wellesbourne, Warwickshire, UK

Multiple births

How is it that birds which lay a large number of eggs are able to have them all hatch on approximately the same day?

Patrick Casement
London, UK

In the majority of bird species synchronous hatching is the norm. This is achieved simply: the parent does not commence incubation until the clutch of eggs, however large, is complete.

Most birds will lay an egg each day until an appropriate trigger indicates that the clutch is complete. This is why I can collect an egg a day from each of my hens – they keep laying until incubation is triggered by some stimulus. This could be the feel of a full clutch against the brood patch on a bird's belly, but there is also some endogenous control factor. For example, one of my hens may turn 'broody' – she will cease laying and commence incubation on only one egg if that is all I have left her with.

In some other groups of birds – notably owls, raptors and cormorants – incubation commences when the first egg is laid, leading to sequential hatching, with the first chick gaining a significant advantage over later ones. This is very noticeable in barn owls, where five or six eggs laid at daily intervals lead to the oldest chicks being almost a week older than the youngest. This strategy ensures maximum chick survival in species with an unpredictable food source. In years of plenty all the offspring get enough food, but in years when food is less readily available the oldest, largest chicks survive and dominate the younger ones – which almost inevitably perish and may be eaten by their siblings.

This strategy sees its most elegant expression in the Cain-and-Abel syndrome, which is manifested particularly well in eagles. The first egg hatches two to three days before the second and, when food is scarce, there seems to be a degree of inevitability in the way the older chick persecutes its younger sibling to the point of death.

Norman McCanch
Canterbury, Kent, UK

Grrrr

How long can you keep a tiger cub as a pet? I have read of people doing so, but surely, for very obvious reasons, there is a time limit to how long you can keep a carnivore in your living room.

Peter Higginson
Paris, France

You can keep a tiger cub as a pet until it grows up and gets hungry or loses patience with you.

Doug Grigg
Cannonvale, Queensland, Australia

If there is an upper age limit then one is assuming that tigers are suitable for a domestic environment below that age. The flaw with this is that they would need their mother at this stage and, if one has qualms about keeping an adult tiger, then one would not even contemplate a maternal tigress.

As for the implication that there is a stage in a juvenile tiger's physical development which would mark a watershed in pet/owner relations, then it is a case of 'take your pick'. Cubs become fully mobile at about eight weeks, when they are still endowed with cuteness, but you wouldn't want one to bite or scratch you. At 18 months, young tigers become independent in terms of being capable of fending for themselves. Fending for themselves in this context generally means hunting. That does not mean they have left home, however. In the wild, cubs stay with mothers for up to two and a half years, and removing one may have emotional repercussions for an animal that is already a dangerous predator.

Even if you and your family avoid featuring on the tiger's menu, there are the minor social, and, probably, legal difficulties arising out of the disappearance of your neighbours' pets. Tigers do not only kill to eat; in the wild they will also suppress local populations of any rival carnivores in their territory. These include wolves – and next-door's labrador. And, if it's a male tiger, there's the particularly noxious spray-marking of personal space to consider – tiger pee makes fox urine smell like Chanel No. 5.

Clearly it is not impossible to keep a tiger as a pet. According to the most recent statistics from the US Association of Zoos and Aquariums, some 12,000 are kept as pets in the US alone, ironically more than the world population of wild tigers, and largely as a consequence of overbreeding by zoos in the 1980s and 1990s. How many are kept as domestic pets in the accepted sense, rather than in wildlife parks or private enclosures, isn't known.

However, in a sad commentary on US animal welfare legislation, a large number of these big cats must be living in people's homes, as if they were large dogs. For instance, the American Society for the Prevention of Cruelty to Animals estimates there are 500 lions and tigers in metropolitan Houston alone. While 19 states have banned private ownership of big cats, only 15 of the others require an owner's licence, and 16 have no relevant legislation at all, despite the US's indigenous population of cougars being on the endangered species list.

These animals are all adults, so strictly speaking there is no upper age limit for keeping a tiger as a pet because it has grown too large, or its behaviour makes it unsuitable. The correct answer to the question is, of course, that no tiger of any age should live socially alongside people. Their proper place is in the wild.

Hadrian Jeffs
Norwich, Norfolk, UK

The power of three

None of the countless species of animal in existence has three legs. Creatures such as the kangaroo and the meerkat use their tails for balance, but a tail is plainly not the same as a leg. This pattern does not apply only to mammals – other kinds of animal have an even number of legs, too. Why wouldn't having three legs work?

Monika Hofman
London, UK

A tripod is wonderfully stable, so there could be something to be said for having three legs. When insects walk, they use their legs as two sets of three. At any instant their weight is supported by three legs – two on one side of the body and one on the other. Meanwhile, the other three legs can be moved forward to form the next 'tripod'.

All the animals mentioned in the question are bilaterally symmetrical, so it is not surprising that their limbs come in pairs – two in the case of land-dwelling mammals, three in insects, four in spiders, and various larger numbers in crustaceans, centipedes and millipedes.

In contrast, starfish are built on a radially symmetrical plan (also seen in sea urchins and sea cucumbers), so they often have five arms. However, these are not like legs, in that they are not manipulated for locomotion. Starfish move using thousands of hydraulically operated tube feet, arranged along the undersides of their arms.

If you had to walk on exactly three legs – as opposed to the insect's two sets of three – you would not want an asymmetrical gait with two legs on the left and one on the right, or vice versa. But an arrangement with one leg on the midline and one on each side is certainly feasible. Having recently been getting about on one leg and a pair of crutches, I can confirm that you can move quite quickly this way, though it is tiring and more difficult on slopes and steps than using two legs.

I think we have to conclude that three legs is an unlikely arrangement in a bilaterally symmetrical animal, and seems to confer no advantage in movement over two or four.

John Gee
Aberystwyth, Ceredigion, UK

As a long-term user of crutches I walk with three 'legs' as often as not. Quite a few gaits are possible while your weight is borne by two legs and crutches, but if you have just one weight-bearing leg you are forced to move the paired outer 'legs' (the crutches) first, followed by the one in the centre (your real leg). The only latitude is in whether you move your leg just as far forward as the crutches, or past them.

Walking with crutches uses up energy at a rate that is typically closer to that of running than walking, indicating that the use of crutches is not an especially energy-efficient way of getting about. Of course, unlike real legs, crutches do not have joints and elastic tissues that can store and release energy to optimise their efficiency, so the potential to evolve an efficient gait using three legs may well exist.

David Gillo
Chatham, Kent, UK

Kangaroos have strong tails capable of bearing weight, and though they do not have any 'three-legged' gaits, they can move slowly with a 'five-legged' gait. First the tail and forelegs are used to support the animal while the hind legs are brought forward in unison, then the hind legs take the weight while the kangaroo shifts forward before putting its forelegs and tail back onto the ground. Because the forelegs are short, the head stays close to the ground throughout, making this gait good for grazing.

The first vertebrates to walk evolved from fish, which swim with a lateral motion, so the gait they evolved probably also involved side-to-side movement. If fish had evolved differently, swimming with a vertical tail motion like a dolphin, then the first vertebrates would have had a gait with some up-and-down motion, possibly using the tail as a 'leg'. In this alternate reality, a five-legged gait similar to a grazing kangaroo could have been common, and tripedal creatures could conceivably have evolved.

Stuart Henderson
Farrer, ACT, Australia

Whispering trees

Different kinds of tree make different sounds when rustling in a summer breeze. What is the physics behind this?

Robin Trew
London, UK

Any airflow disturbance, such as that caused by leaves, creates sounds of characteristic volume, frequency and oscillation. Trees' songs change with wind speed and direction, and the type of leaves.

Needle-like leaves shed vortices as the wind oscillates round them, creating the high-pitched, romantic whisper of conifers. Flat leaves flap like flags, depending on thickness, firmness, edge outline and surface texture. This is commonly the main component of the rustling sound. Pointed, narrow willow leaves shed wind energy with whisperings.

Colliding leaves suffer damage, so they grow in patterns to avoid touch. In high winds, though, impact is inevitable, causing another kind of rustling. Smooth, large, simple leaves tend to give low notes except when flapping vigorously. Trees with small leaves, prominent veins, complex outlines, furry surfaces and rough bark seem quieter, but produce ultrasonic sounds.

Crisp autumn leaves act as rattles. Hollow leaves emptied by aphids, and acacia thorns hollowed by ants, may whistle. Dense foliage dampens high notes. Leaves on high branches differ in shape and texture, and encounter higher winds. The leaves of rushes scrape and vibrate like the reeds of wind instruments, giving rise to the Greek legend about their whispering: 'King Midas has ass's ears!'

Jon Richfield
Somerset West, South Africa

As the wind blows it causes leaves to strike other leaves or twigs, which creates percussive sounds, and to flex or deform. As the leaves deform, energy is released which is dissipated as sound in a similar way to how sound is created when wrapping paper is crumpled or flattened out.

Apart from the speed of the wind, important factors determining the type of sound created include the stiffness of the stalks, proximity of a leaf to its neighbours, and the size, stiffness, weight, moisture content and shape of the leaves themselves. The impact of large, heavy leaves will be louder than that of small, light ones, in the same way that the sound created when you tap a postcard on a table is louder than if you tap with a piece of paper.

Dry leaves are stiff and brittle so more energy is required to flex them, which means that more energy is available for dissipation. Therefore the rustling of the dry, desiccated leaves of a beech hedge in spring will sound quite different to the quieter sound of its pliant summer leaves.

Richard Holroyd
Cambridge, UK

The colour of anemone

*For a scuba diver, one of the best underwater sights is a rock face covered in brilliantly coloured jewel anemones (*Corynactis viridis*). They exist in many colours, and often vivid contrasting colours are found side by side. There are also subdued, semi-transparent variants. Most species of wild animals have evolved to just one or a narrow range of colours, while flowers can have a range of vivid colours, presumably to attract a variety of insects. As far as I know, the anemones aren't trying to attract their prey – it just arrives on the current. So why are they so vivid and so varied?*

George Gall
Truro, Cornwall, UK

My colleague Anya Salih and I have worked on this question for some time in corals, which are close relatives of sea anemones. We believe that the pigments have a protective function against excess light, as discussed in our paper 'Fluorescent pigments in corals are photoprotective', which appeared in *Nature* (vol. 408, p. 850).

Unpigmented as well as pigmented versions exist in both corals and anemones. The explanation for this is probably that the production of pigments is 'costly', and pigmented versions cope by being fitter than their unpigmented cousins. When conditions are unfavourable the coloured ones do better, though favourable and poor conditions are both common enough that neither form takes over.

Some controversy over this interpretation remains, although we are still waiting for someone to come up with something better.

Guy Cox
Electron Microscope Unit
University of Sydney
New South Wales, Australia

Three's a crowd

In a clumsy effort to seduce her, I was trying to explain the evolutionary advantages of sexual reproduction to a female friend the other day, one of which I said was introducing an element of genetic competition into the process. She wanted to know why, if two sexes are needed to create genetic competition, there aren't three, four or a million sexes to create even more competition. Why are there only two?

Tim Rowland
Bristol, UK

Some species do have more than two types. Single-celled ciliates have up to 100, and mushrooms have tens of thousands. But most organisms – even single-celled ones – come in two types.

So why are there two types in most species? In all species, no matter how many types, sex occurs between just two cells and any can mate with any other sex cell that is different from it. So, as your questioner suggests, finding just two types in most species is paradoxical, because having many types would maximise the chances of finding a mate.

One answer to this problem is that two types is best for coordinating the inheritance of cytoplasmic DNA – the part of the cell's genetic material that is not contained in the nucleus. However, there is a drawback to this solution. The species with two types fuse cells and potentially run the risk of scrambling this extra material.

The species with more than two mating types do it differently. With three types, the coordination is even more difficult to make error-proof, while those with many mating types don't fuse cells at all and so are not constrained to having just two types.

Laurence Hurst
Professor of Evolutionary Genetics
University of Bath
Somerset, UK

Laurence Hurst has written widely on the subject and more information can be found in his following papers: 'Cytoplasmic fusion and the nature of sexes' (with William Hamilton), Proceedings of the Royal Society B, *vol. 247, p. 189; 'Selfish genetic elements and their role in evolution: the evolution of sex and some of what that entails,'* Philosophical Transactions of the Royal Society B, *vol. 349, p. 321; 'Why are there only two sexes?'* Proceedings of the Royal Society B, *vol. 263, p. 415 – Ed.*

Having taught a difficult lesson on statistical techniques in geography to my secondary school students, I stood before them lost in admiration of the chi square test I'd written up on the board. Just then my students informed me that there were actually three sexes in this world: men, women and geography teachers. Unfortunately, I am a geography teacher.

Mary Sinclair
By email, no address supplied

Iron plants

Iron deficiency is common among human vegetarians, so how do herbivores cope?

Melanie Green
Hemel Hempstead, Hertfordshire, UK

Vegetarians have dietary difficulties because they force their omnivorous physiology to cope with a herbivorous diet, mineral imbalances being only one of the consequences.

Herbivores survive in good health partly because some are not as vegan as we might imagine. They eagerly eat animal dung, old bones, incidental insects and the like. They are also not too proud to eat dirt wherever they find a salt lick. Also, practically all herbivores rely on a partnership with gut flora to supply micronutrients or improve digestion.

Then again, they need to eat huge volumes of vegetation to ensure they absorb sufficient quantities of minerals from the minute concentrations in plants. After all, plants contain a little iron and manganese as well as macronutrients such as magnesium because these are needed for photosynthesis.

Humans trying to match the performance of specialist herbivores would need bellies like proboscis monkeys, and would be eating 18 hours a day just to keep up; never mind the consequent activity at the nether end, nor the tooth wear that, as brachydont herbivores, humans would suffer.

Jon Richfield
Somerset West, South Africa

? The anthill mob

On a recent summer trip to Ushuaia in Patagonia, Argentina, we saw no ants. This troubled us so much that we ended up actively searching them out, with no success. Is there a southern – and indeed a northern – limit to the range of ants or were we just looking in the wrong places?

Andrew and Bronwyn Lumsden
Murrays Run, New South Wales, Australia

In searching around Ushuaia, the questioners found one of the few places on land where ants do not occur naturally, although there is the possibility that adventive species – that is, non-native and non-established ants – survive in houses. It's just too cold and wet there. Other places where you can look in vain are Antarctica, the sub-Antarctic islands (although an adventive species was found once in an abandoned whaler's hut on Kerguelen), the Falkland Islands, the high Arctic, Iceland and the upper slopes of high mountains.

Edward O. Wilson
Department of Entomology
Harvard University
Cambridge, Massachusetts, US

Foxy tale

From a slow-moving train I saw a fox standing with its tail resting on the ground while two magpies repeatedly took turns to peck the tip of the tail, before running off. The fox merely flicked its tail each time. What were they all doing?

Sue Murdochs
Marton cum Grafton, North Yorkshire, UK

Although counter-intuitive, the magpies that you observed may have been trying to prevent themselves becoming the fox's next meal. Animals have evolved a variety of ways to avoid predation. One strategy used by a number of species is to advertise that you are so fit and healthy that it would be pointless for the predator to waste its time trying to catch you. The magpies may have been doing just that.

Alternatively, if you saw this behaviour in the breeding season, the magpies may have been trying to protect their young. Many bird species will do everything possible to prevent predators from approaching their nesting sites or their newly fledged chicks. Birds achieve this either by acting as bait in order to lure the predators away from their offspring, or by directly attacking predators to drive them away. The ferocity with which birds attack predators often increases as the predator approaches the nest.

Magpies can become very aggressive towards intruders, so it is likely that the fox that you saw wasn't close enough to be a real threat, but was being given a friendly warning.

John Skelhorn
The Institute of Neuroscience
Newcastle University, UK

I have a magpie nest in my garden. Every spring, when the chicks hatch, both parents harass my two cats to prevent them approaching the tree. Sometimes the cats can hardly get out of the house. These magpies are quite fearless, chasing the cats right to my doorstep, and they keep up a loud and disagreeable chatter.

This behaviour lasts about a month, but can also happen in autumn or winter, even though the nest is empty. The magpies seem only to attack potential predators in pairs, but other members of the crow family are known to gang up against predators. I've seen flocks of crows attacking a buzzard or an owl.

Annelise Roman
By email, no address supplied

I have rescued sick and orphaned wildlife for 39 years, specialising in rooks for the past decade, and have raised and released more than 200. Magpies and rooks are members of the crow family, and I have often seen the behaviour mentioned.

My dog is tolerant enough to let juvenile rooks ride on his back, and they will tease him into interacting with them. Some of the really mischievous characters among them play exactly the same game observed by the questioner. When my dog is napping they will tug at the end of his tail or ears, and jump out of reach when he reacts by flicking his tail or by shaking his head.

Magpies take up to three years to mature, and I have five

non-flying resident adults to act as tutors. Without them I would be releasing a crowd of juvenile delinquents back into the wild.

The game of tail-tweaking described above was possibly being performed by young magpies and, if they are like my rook babies, they were probably doing it for fun. Among their many games, this is a favourite.

Tina Kirk
Swaffham, Norfolk, UK

Lord of the flies

During summer, if I drive my car in Europe the windscreen is splattered by dead insects. However, I can drive for months in the Caribbean without any insects striking the screen. Are Caribbean insects smarter?

Erik Blommestein
Trinidad and Tobago

It isn't a matter of Caribbean bugs being smarter but of what the local conditions are like or even something as simple as the road size. If you drove along a minor country road in Europe or along an urban highway in the Caribbean, for example, you would probably experience windscreen splatter of the kind you describe.

Many insects are drawn during the evening to any source of light, even if it is only faint, and so may fly towards car windscreens. Insects would not be similarly enticed if you were driving at noon. So the number of insects hitting your screen could be affected by different light levels in Europe and the Caribbean. To make a sensible comparison we need more details, including which species perished.

Jin Xiao
National University of Defense Technology
Changsha, China

The effect could be a consequence of differences in road infrastructure in Europe and the Caribbean. European roads are mostly wide and asphalted, and allow high-speed driving, whereas Caribbean roads are not always so welcoming to the fast driver.

In Europe, traffic and asphalted roads generate and retain heat, which is usually appreciated by insects. In the Caribbean, the difference in temperature on and off-road is not so great, meaning the road environment may be less attractive to insects. Also, you would probably have to drive more slowly in the Caribbean, giving the insects a chance to escape or be swept over the car by the airflow.

Gael Canal
Chaville, France

You haven't experienced the full fury of bug splatter unless you have driven in Florida in early spring or late summer. The culprit is *Plecia nearctica*, the love bug. The name is well chosen because the insects fly together in pairs. The male locks the end of his abdomen with that of the female, and when she starts to fly the male arches upside down over her and flies with her. In a swarm you seldom see a love bug without a partner flying upside down above it, though this arrangement makes for a very low flying speed.

Love bugs can swarm above roads in such numbers that they almost form a cloud. When you hit an infested section of highway, it can completely obscure your windscreen in one quick blast, as well as the radiator and headlamps. If the black remains are not removed from the paintwork, they can corrode the finish in a day or two.

At the height of the love bug season, many drivers fix a screen to the front of their car to make cleaning them off easier. In the north of Florida, washing stations have been installed along the main highways so drivers can clean the bugs off their vehicles.

William Joseph
UK

Family tree

What, if any, are the visible inherited traits of trees? People often say a child has its mother's or father's eyes. Does a similar thing apply to trees? For example, is the pattern of branches related to the position and orientation of its parents' branches? If not, what governs where branches grow?

Graham Cook
Newcastle-under-Lyme, Staffordshire, UK

The anatomical, chemical and morphological characteristics found in trees are under varying degrees of genetic and environmental control, just like the traits of other organisms. For a given characteristic, the proportion of total variation that is explained by genetic control is known as its heritability. This can range from 0 to 100 per cent: the higher the value, the more closely progeny resemble their parents and the greater the improvement that can be obtained by selective breeding.

The genetic variation itself includes additive, dominance and epistatic effects, in which one gene modifies the action of another. The proportion of these effects varies between species, populations, environments, plantation management conditions and also between the individual trees. Some pairs of genetically related traits interfere with each other, making it tricky to obtain benefits in both by selective breeding.

Two major sets of visible traits are under a high degree of genetic control and are of great importance for the end use of the wood. One is 'stem form', which is a measure of sinuosity or deviation from perfect straightness. It is associated with 'reaction wood', which forms in response to mechanical stress, such as exposure to strong wind. Reaction wood is visibly asymmetric and is generally undesirable for end uses such as solid wood and pulping for paper.

The variable characteristics of the tree's branches are also important to industry and include the thickness, number per unit of stem length, angle of insertion on the stem, and whether branches occur in whorls or are scattered at random along the length of the stem. These variables affect the number and size of undesirable knots.

Other characteristics in which parents and progeny may be similar include flower colour in horticultural species and flower or fruit colour, shape, size, flavour and nutrient content in fruit trees.

Jeffery Burley
Director Emeritus
Oxford Forestry Institute
Abingdon, Oxfordshire, UK

Saucy dogs

I have two female dogs whose urine kills the grass in patches all over the lawn. My mother advised feeding them tomato ketchup, which I did, and the patches stopped appearing. Why does this work, and should I really be feeding my dogs tomato ketchup?

Jim Landon
Swindon, Wiltshire, UK

The urine acts as a liquid fertiliser, but can produce nitrogen overload where the puddle of urine is deepest. This 'burns' the grass, creating a brown patch in the lawn.

Towards the outside of the puddle, where less nitrogen has been applied, there can be a fertilising effect leading to a ring of luxuriant, greener grass. The urine of dogs and bitches does not differ much but, while dogs tend to deliver small samples of urine to mark their territory, bitches tend to empty their bladders entirely, causing more harm.

Urine is slightly acidic, but so is tomato ketchup, so it does not neutralise the urine as some people believe. Instead, the salt content of tomato ketchup, juice or sauce makes dogs drink more, diluting the nitrogen in their urine.

Be aware that increased salt intake can cause problems with existing kidney or heart conditions, so if you must tinker with your dogs' diet, consider reducing the protein content instead. This will also reduce the nitrogen content of their urine, and should be fine for all but the most active of dogs. Better still would be to train your dogs to urinate in a designated place or follow them out of the house with a hose pipe or watering can to dilute their urine.

Mike Follows
Willenhall, West Midlands, UK

Sight or flight?

Butterflies have a very haphazard flight pattern that no doubt serves them well as a defence against predators. But with their small eyes and small brains how can they see or understand where they are going?

Peter Koch
Le Touvet, France

Release a nocturnal moth by day and as likely as not a bird will gobble it up. But birds normally take no interest in daytime butterflies, which move about freely, seemingly ignored. Nonetheless, I have seen birds try to catch butterflies on several occasions and each time it was clear why they are normally left alone: they are too quick and manoeuvrable.

It is not true, though, to say butterflies have small eyes. Each compound eye covers most of the side of the head and is huge compared with the size of the insect. They evolved compound eyes because a simple eye with the same light-gathering power would be heavy and cumbersome.

Butterflies also have three small simple eyes, called ocelli, on the top of the head. These serve, so far as is known, at least two purposes. They help the insect maintain a correct flight attitude and they assist in navigation. All day-flying insects have large eyes compared with their nocturnal counterparts, showing the relative importance of vision.

Typically, although not exclusively, a male butterfly will stake out a territory in a sunny spot and display to attract a female, seeing off any other male with the same idea. Initially at least, the female finds a partner by sight, observing the brightly coloured display. Butterflies also find flowers by sight to feed on their nectar.

They do not see the same light spectrum as humans: theirs is shifted into the ultraviolet, while red is invisible, so flowers usually have a UV component.

Humans cannot see the UV reflected from a field of oilseed rape in flower on a bright day, but we can tell it is there from the shimmering, dazzling character of the light.

As for understanding, there is not much room in a butterfly for a brain. The nervous system consists of a series of connected ganglia – collections of nerve cells – with the ganglion in the head merely larger than the others.

Priority has to be given to motor control and reflex behaviour; there is no space for much more. To survive, butterflies have to do four things: feed, mate, find a plant on which to lay eggs and, in certain cases, migrate. Understanding is not necessary.

It should not be forgotten either that butterflies have antennae – quite a lot of 'seeing' is done with these. A male will not be deceived by a predator simulating a female because the true female gives off a pheromone sensed by receptors on

the antennae. Flowers, carrion, faeces and rotting fruit can all be found by odour too. This is also how a female finds the correct plant on which to lay eggs. The butterfly confirms the plant's identity by tasting it with her feet when she lands.

Terence Hollingworth
Blagnac, France

Buzz off

Why do flies that enter your house in summer do so in such an ostentatious way, with loud and frenetic buzzing? Surely it would be better for their survival if they didn't? Is there a purpose to this display?

Greta Bowman
Brighton, East Sussex, UK

The noisiest flies are blowflies, especially the bluebottle *Calliphora vicina*, which beats its wings around 150 times per second. This generates air vibrations equivalent to 150 hertz; musically speaking it is a D, below middle C. They are noisier than the housefly, *Musca domestica*, because their bigger wings disturb more air. Flies sound louder indoors because there is less background noise to drown them out and their buzzing may reflect off solid surfaces. Their buzzes sound more mechanical than musical because they lack the sweeter harmonics.

Flies clearly evolved with peskiness in mind – forgive my anthropomorphism. They enter without knocking. They aim at your head for their fly-by. They vomit on windows, and are impossible to swat. Then to be really cussed, they taunt you by loitering out of reach and parading about your ceiling. But we might not be asking this question were our hearing beyond their beat frequencies.

The buzz is part and parcel of the very wing beats that initiate flight, so it would be impossible for flies to dispense with buzzing without having to adopt different aerodynamics.

Len Winokur
Leeds, UK

Flies forage constantly, mostly by smell, so they search an area thoroughly before moving on. They treat a house as they would any enclosed area, such as a cave, and when they have finished they look for a way out by using light intensity. In homes this often brings them into contact with glass, which confuses them because they have not evolved to understand it. What amazes me is that a fly can fly full-tilt into a pane of glass without apparently injuring itself.

Since their predators, mostly reptiles and birds (and sometimes pet dogs), tend to hunt by sight and not sound, flies don't care how much noise they make.

Tony Holkham
Boncath, Pembrokeshire, UK

Some flies, such as the irritating blackflies of the family Simuliidae, which buzz around your head outdoors on still days, thankfully do not enter buildings. Houseflies such as *Musca domestica*, blowflies such as *Calliphora vomitaria*, and greenbottles of various *Lucilia* species, have no such scruples.

Most people do not realise, I imagine, that much of insect behaviour is regulated by odours. A mosquito finds its victim primarily by detecting body odours and chemicals in breath. The flies described are similarly attracted, not to people, but to chemical signals from food. The blowflies and greenbottles coming into my kitchen make straight for the food remnants in my cat's dish.

If I forget to put it in the fridge during warm weather – the dish, not the cat – it is soon speckled with white patches of

eggs. Since the primary food source of such flies is carrion it is hardly surprising that they also do not hesitate to enter dark spaces where a sick animal may have taken shelter and died.

During the colder months the same flies will seek out a sheltered corner indoors where they can hibernate, perhaps behind that piece of peeling wallpaper in the top corner of the room. If a cold house or room warms up, the heat often spurs hibernating flies into activity, although they often then fly around in a drunken, sluggish manner.

I have on occasion shooed flies out of the open house door into the cold only to see them realise their mistake and make a smart U-turn back into the warm.

A house has no significance to a fly, other than as a source of heat, shelter and food. Humans have been around a few hundred thousand years at most while insects have been on Earth at least 400 million years. A few flies trapped or swatted inside houses are irrelevant to their survival as a species.

Terence Hollingworth
Blagnac, France

Growth spurt

What factors determine the point on the surface of an apparently homogeneous potato at which a new sprout bursts forth?

Peter Foley
Workington, Cumbria, UK

The clue is in the word 'apparently'. A potato tuber is not just an amorphous lump of dinner, but a modified stem. 'Stem' in botany is a technical term – it is a structure that has polarity (an apex and a base), grows leaves out of its sides, and has a bud in the angle, or axil, between the stem and each leaf.

The tuber differs from a typical stem in many respects. First, it originates as a swelling at the tip of another special stem, called a stolon, which is itself unusual in that it grows downwards into the soil instead of upwards into the air, and which later withers, leaving a scar.

Second, it is packed with starch, which is why we eat potato tubers, of course, and has only vestigial leaves – nothing more than slight bumps. But it is unequivocally still a stem, having normal polarity and, importantly, axillary buds, which are initially very small but capable of bursting forth when their hormones give them the urge. These buds can be seen on a mature tuber, and most of the big ones are at the apical end, that is, furthest from the base, where the tuber used to be attached to the stolon. Indeed, most plant stems tend to have more vigorous buds near the apex.

The polarity of a tuber, and of any stem, is hardwired by each cell's innate tendency to pump the hormone auxin one way, from apex to base. Potato tubers also have a thin outer layer of cork – the 'jacket' – which is pitted with lenticels, or breathing pores, like the bark of a tree trunk, again emphasising that the tuber is just a tasty stem.

Stephen C. Fry
University of Edinburgh, UK

Potatoes are stem tubers: they have all the features of a stem, such as a terminal bud, lenticels, and internal vascular bundles arranged in the stem pattern. They also have axillary buds, which are commonly known as the 'eyes'.

It is from these buds that the shoots, or haulms, grow. You can encourage them by leaving potatoes in the light, which is called chitting. Potato growing, or 'solanoculture', has its own rich terminology, which I have barely scratched.

Luce Gilmore
Cambridge, UK

Language classes

Will humans ever be able to speak or even understand dolphin?

Riccardo Pesci
Rome, Italy

The idea that dolphins possess a communication system as sophisticated as human language was proposed by John Lilly in the 1960s. Despite loud protests from a sceptical scientific community, Lilly vowed that pioneering researchers would one day 'crack the dolphin code' and begin an interspecies dialogue.

In the ensuing years, dolphins were taught to use artificial symbol systems, with equivocal results. Their performance is comparable to great apes where comprehension is concerned, but when it comes to using symbols to establish two-way communication with humans, dolphins have been overshadowed by linguistic prodigies such as Kanzi the bonobo.

Dolphins' own communication system has been the subject of much study, revealing a perplexing array of vocal and non-vocal signals. But a sober view of half a century's worth of evidence suggests that dolphin communication – even when taking into account the referential, or word-like, nature of their mysterious 'signature whistle' – is nothing more than a variation on the type of communication system seen throughout the animal world. It is complex, to be sure, though likely to be short on content. There is little to suggest that the cacophony of whistles and buzzes is used to share limitless, abstract information in a language-like fashion.

Science is destined to make great strides in unravelling the mysteries of dolphin communication, as there is much we do not yet understand about the function of their vocalisations. However, the idea that dolphins are harbouring a secret language that awaits decryption is looking increasingly like a spot of wishful thinking from a bygone era.

Justin Gregg
Research Associate and Vice-President
Dolphin Communication Project
Old Mystic, Connecticut, US

Dolphins and humans can communicate, but is it possible for them to engage in meaningful conversation? Perhaps, but communication between the two species has been limited to date. In fact, there is no compelling scientific evidence that humans and dolphins can engage in exchanges of information beyond those that involve a human requesting a dolphin to perform some behaviour or those that inform a human about some object the dolphin would like but cannot obtain without human assistance.

There are many possible reasons why we cannot converse with dolphins, including the fact that we have much to learn about their communication systems. Dolphins produce a variety of sounds and other behavioural cues that appear to be communicatively significant, yet the communication units used by dolphins remain unknown. For example, are whistles separate single units or some combination of smaller units? Once the units that comprise the dolphin communication system are ascertained, the daunting task of determining what they mean remains. This will require a comparison of how individual units are used in isolation and with other units in a variety of contexts, a process that has only just begun.

Clearly, such work is necessary before conversations can occur. However, it may not be sufficient. Conversations require shared interest in a topic, so humans will need to find subjects that interest dolphins. Given the differences between us, discovering a common ground for meaningful conversations may be more difficult than some humans imagine.

Stan Kuczaj
Marine Mammal Behavior and Cognition Laboratory
University of Southern Mississippi
Hattiesburg, Mississippi, US

Wall-to-wall web

In a chalet in the Alps which had no obvious draughts, I noticed spider threads spanning horizontally from wall to wall. How did they do it?

Marcello Rebora
London, UK

Air currents carry a silk line from a spider's spinneret until it reaches a solid object. Because of its sticky nature, the line attaches to whatever it encounters. Draughts or convection currents sufficient to carry the line would exist even in a room that was hermetically sealed. For example, windows allow sunlight to penetrate, heating a patch of floor. Air in contact with the floor would warm and expand. Being less dense, it would rise, to be replaced by cooler, denser air, initiating the required air movement.

The spider scurries along the first line – called a bridge line – spinning a stronger second thread. It continues making return journeys until the line is sufficiently strong. The rest of the web follows.

David Attenborough describes the construction of a web by an orb-web spider at bit.ly/fC8ABR.

Mike Follows
Willenhall, West Midlands, UK

4 Shaken or stirred?

Stirring stuff

What is the significance of James Bond's famous phrase 'shaken, not stirred'? Is there really a difference in the taste of a shaken vodka martini, as opposed to a stirred one? And if there is, why?

Mark Langford
Stockport, Cheshire, UK

The shaken-versus-stirred dispute has run for years in The Last Word. One of our earlier books, Why Can't Elephants Jump?, *provided what we genuinely believed was the last word on 007's favourite tipple. How wrong we were. Sceptical readers continued shaking, stirring, freezing and drinking in a determined effort to get to the bottom of the martini mystery. Readers with elephant-class memories will be familiar with the first few answers – but bear with us: all research must start somewhere and those short, simple answers triggered an unquenchable thirst for knowledge and the development of the new field of martini science. Months of selfless devotion to the cause by our international team of martiniologists has brought results that have left us drunk with knowledge on this vitally important subject.*

Supposedly, when a martini is shaken, not stirred, it 'bruises' the spirit. To seasoned martini drinkers this changes the taste.

Padraic O'Neile
Newcastle, New South Wales, Australia

Because a martini is to be drunk within seconds of preparation rather than minutes, there is a difference. The tiny bubbles caused by shaking mean a well-shaken martini is cloudy. Shaking will also have an effect on the drink's texture – making it less oily than the stirred version – and hence on the taste. The long-standing assumption that the spirit is bruised by the process is nonsense; vodka does not have a vascular system.

Peter Brooks
Bristol, UK

Peter Brooks was the first correspondent to mention oily textures found in stirred martinis which hints at the correct answer to the original question – Ed.

Bond may have appreciated the softening and ripening effect of partial oxidation of the aldehydes in vermouth – akin to letting red wine breathe before you drink it. In a refined and homogeneous substrate such as vodka martini, a good shake can speed up the process.

Alan Calverd
Bishop's Stortford, Hertfordshire, UK

Canadian biochemists, however, had other ideas on the subject – Ed.

Biochemists at the University of Western Ontario in London, Canada, have suggested the change in flavour brought about by shaking is due not to the oxidation of aldehydes, but to the breakdown of hydrogen peroxide. Stirred martinis have double the amount of hydrogen peroxide as shaken ones.

Peter McNally
Vancouver, British Columbia, Canada

The reason shaken martinis are cloudy is not so much down to bubbles, but because the crushed ice in the shaker deposits tiny ice crystals into the poured drink. The drink slowly clears as the crystals melt.

Frank Melly
New York City, US

At this point, we felt duty-bound to do some experiments of our own. Was it bubbles or ice that caused the cloudiness in a shaken martini? And could either account for any difference in taste?

First, the martinis. We borrowed our recipe from cocktail mixologist Eric Keitt who works the bar at P. J. Clarke's in Washington DC:

Double vodka and a few drops dry vermouth
Pour into a cocktail shaker with crushed ice
Shake until the hand holding the shaker is very cold
Strain into a martini glass
Add an olive or a twist of lime zest

Keitt also provided this advice: 'The application of vermouth should be a few – and I mean a few – drops, maybe two or three. Vermouth will release the aromatics of the vodka, making for a more enjoyable drink.' Keitt is a fan of stirring the drink for reasons given below, but in this instance we had no choice but to shake.

Three martinis were prepared. The first was shaken with crushed ice. The drink was very cloudy and took a long time to clear but, as far as we could ascertain, the cloudiness was caused only by tiny bubbles produced by the shaking plus the condensation on the chilled martini glass. There were no ice crystals that we could see.

The second was a room-temperature martini, shaken without ice. Bubbles formed in this when it was poured but quickly dissipated, much faster than in the iced martini.

The third martini was made in an attempt to replicate the chilled conditions of the iced martini but without adding ice to the

shaker. The martini and its shaker were wrapped in a drinks chiller until ice cold and the same temperature as the first martini. Then it was shaken. When poured it was cloudy for much longer than the room-temperature martini but not for as long as the iced martini.

From this we ascertained that the ice does have some effect on the clouding process, as do cold conditions. Iced martinis do produce the cloudiest drink, but no ice spicules appeared present in the drink contrary to the earlier suggestion by Frank Melly. Chilled martinis without ice produced a cloudy drink too, but for a shorter time than iced martinis. The room-temperature martini cleared the fastest. So nothing conclusive here yet, except that temperature plays a role of some kind. More experimentation was needed, so we asked if there was any reader out there who could examine the shaken martini microscopically to rule out (or, indeed, confirm) the presence of ice crystals?

Putting cloudiness to one side for the moment, we also heard from Anna Collins, who seems to have answered the original question of what makes a shaken martini taste different from a stirred one, her suggestion apparently being confirmed in a blind tasting – Ed.

The reason Bond orders his martinis shaken is that the ice helps to dissipate any residual oil left over from potatoes – the base ingredient for many vodkas at the time Ian Fleming's novels were written. With the rise of higher-quality grain vodkas, shaking is unnecessary, and for many fans of the vodka martini, shaking the drink with ice dilutes it too much. Stirring with ice chills it without reducing its strength.

Anna Collins
Washington DC, US

So now we had to get down and dirty with potato vodka. Fortunately one reader decided to check out whether this really was the case – Ed.

Anna Collins is correct, according to our blind trial. We bought two bottles of vodka, one grain, the other potato-based. First we tasted the vodkas. In the blind trial all six people in our sample said the potato vodka was oily, and the grain vodka wasn't. Then we made two vodka martinis using the potato vodka. One was stirred with ice, the other shaken with ice. The difference was quite distinct and in a blind tasting every one of the six drinkers characterised the shaken martini as being much less oily. But the martini had to be consumed quickly. If left to settle for 5 minutes or so, the shaken martini became oily again.

Peter Simmons
London, UK

The unresolved issue of ice crystals caused some to question our recipes – Ed.

While I would not want to disparage another bartender, I'd dispute the method used in your experiments. Of course you wouldn't see ice crystals if you started with crushed ice. The ice particles are already too small. But if one uses ice cubes – the smaller kind seen in most bars, not the kind you make at home – then the ice crystals are easily visible in a shaken martini. The crashing of the ice against the metal shaker creates these shards that inevitably end up in the drink.

We also need to talk garnishes. The most common are olives or a lemon twist. If shaking vodka is meant to break up residual oils in the vodka, why is the most common garnish oily – olives? Even the rind of a lemon – a twist – has some oil. The variation of martini known as a gimlet, which uses a dash of lime juice, seems the best counter to the oiliness.

Incidentally, here's a bartender's joke: despite writing his novels decades before the martini was 'invented' it's apparent that Charles Dickens supplemented his income by working in

a bar. 'Olive or twist' is a familiar bartending question. Why else would he name one of his characters so?

Molly Navin
Philadelphia, US

And it didn't stop there. Another reader in the UK, William France of Birmingham, wrote in to point out that in the book and movie Casino Royale *Bond orders his martini to the recipe later known as a 'vesper', which contains more gin than vodka, yet he still requests that it is shaken. So why? – Ed.*

Bond's martini in *Casino Royale* is made to the following recipe (with thanks to the 1953 novel by Ian Fleming):

> 'A dry martini,' [Bond] said. 'One. In a deep champagne goblet.'
>
> 'Oui, monsieur.'
>
> 'Just a moment. Three measures of Gordon's [gin], one of vodka, half a measure of Kina Lillet. Shake it very well until it's ice-cold, then add a large thin slice of lemon peel. Got it?'
>
> 'Certainly, monsieur.' The barman seemed pleased with the idea.

We can also presume the vodka used was a potato vodka because Bond goes on to tell the barman: 'Excellent… but if you can get a vodka made with grain instead of potatoes, you will find it still better.'

So we know there is a measure of potato vodka in the original vesper. To check whether this still makes a difference to the martini, I and four friends repeated Peter Simmons's blind tasting of the martinis, making one batch with potato vodka and another with grain vodka. We had to amend the recipe slightly because the Kina Lillet in the vesper (which replaces the dry vermouth used in standard martinis) is no longer produced, the nearest modern-day product being Lillet Blanc. Lillet Blanc is less bitter than Kina Lillet (Kina referring to the bitter quinine that was in the original) so we had to add two drops of bitters to the drink so it matched approximately the original vesper taste.

Without a doubt, the potato vodka vesper was oilier than the grain vodka version (despite containing only a single measure of vodka, as opposed to three of gin). And subsequently, when shaken with ice the oil in the potato-vodka vesper was much less pronounced. All tasters were unanimous in detecting this.

Thus, despite the modern convention of stirring martinis, it seems that Bond, a man of obvious sophistication, knew what he was talking about when always asking for his martinis to be shaken, recipe notwithstanding.

Janice Devaney
London, UK

In the novel *Moonraker* written in 1955 James Bond adds a few grains of black pepper to his vodka (even though it is a grain-based variety) to cause any residual fusel oils to be attracted to and collect around the grains as they sink to the bottom. In the book Bond explains that this is a habit he acquired when posted to the Soviet Union to counteract the negative effects of the illicitly distilled vodka found there which tended to contain high levels of fusel oils.

Fusel oils (also known, ironically in this instance, as potato oils) are alcohols with more than two carbon atoms (unlike the more common ethanol drinking alcohol which has only two). They are produced during the normal fermentation process at higher temperatures or during periods of limited yeast activity often caused by low nitrogen levels in the surrounding gases and they can turn up in alcoholic drinks. During distillation they occur most often in the 'tails' (that part of the distillate that occurs at the end of the distillation process). They are usually discarded because they spoil the taste of the better part of the spirit which is usually drawn from the middle of the distillation process. They certainly have the oily consistency that Bond appears to want to avoid when insisting his vodka martinis are shaken.

Popular belief has it that drinking fusel oils increases the chance of a hangover, but studies into alcohol consumption and its effects have yet to prove this.

Ian Rogers
Milan, Italy

So we'd finally cracked the shaken-not-stirred argument and thought we were done and dusted. But no, there was still a world of martini knowledge to discover – the questions just kept coming – Ed.

For your ice only

To save mixing my gin or vodka martinis with ice and over-diluting them I pre-mix them and put them in my freezer. When I take them out they are a sludgy mixture. But any solid ice that has formed does so in very thin sheets randomly aligned in the sludge which melt quite quickly once the martini is out of the freezer and in a glass. How do these sheets form? I tend to make my martinis about 8 parts vodka or gin to 1 part dry vermouth.

Thomas Park
London, UK

Practically all ice is crystalline; the crystals grow as cold molecules packing onto their outer surfaces, rather like infinitesimal Lego blocks.

When water molecules fit together, certain joints are stronger than others, so the crystals grow faster in some directions than others. Molecules of pure water pack easily in almost any direction, so ice forms nondescript blocks unless freezing is slow. But foreign molecules, such as sugar or ethanol, interfere strongly with packing in certain directions, so needles and plates result.

Such crystals also interfere with each other's growth and create temperature and concentration gradients that favour alignment, leading to the packed crystals found in ice lollies for instance, which grow from the lolly stick that acts as a nucleation site. Ice crystals tend to dissolve in the surrounding liquid, with melting happening largely at the crystal surface; arrays of thin crystals present a large surface, and consequently melt rapidly. This explains why the sheets in the martini melt quickly. The surrounding solution also loses water when the ice freezes, increasing its concentration, and that accelerates melting, much as salt melts snow on roads. The result is a very cold martini, but drink it quickly; it will take up heat correspondingly rapidly.

Jon Richfield
Somerset West, South Africa

Martinis are forever

I've read in many places that when the vodka and vermouth that make up the martini are being stirred with ice you should always use a proper, long and thin bar spoon, or failing that a table knife. This makes a much colder drink than stirring with a normal, wide-bowl spoon. I tested this, stirring two identical martinis for 30 seconds, and found the martini stirred with the knife was an average 2.5 °C while that with a tablespoon was 3.9 °C. I repeated the experiment five times (on different nights, honestly). So why?

Paul Townley
Norwich, Norfolk, UK

As Q might say: 'The effect of the stirring device on the final temperature of the martini depends on how much heat the stirrer adds (or removes) from the mix. Oh, do pay attention 007.'

A stirrer with a large mass will have a bigger potential effect than a small one. This may be one reason why 'proper' bar spoons are long and thin, although I suspect it may also be to reach the bottom of tall glasses or cocktail shakers. A tablespoon would certainly have a bigger warming effect than a knife, assuming both implements initially were at the same temperature and warmer than the drink.

The rate of heat transfer from stirring implement to drink depends on the thermal conductivity of the implement. Thermal conductivity varies quite widely between metals; a solid silver spoon would conduct heat about 16 times as fast as a stainless steel spoon. Heat would be conducted from the mass of spoon immersed in the drink, but it would also flow from the shaft and thus from the barman's hand. This effect would be smaller the thinner and longer the shaft, which may be another reason for the shape of bar spoons.

The amount of heat available for transfer from the spoon to the liquid depends on how much warmer the spoon is than the liquid and on the specific heat capacity of the spoon. For metals, specific heat capacity does not vary quite as much as thermal conductivity – it's about twice as high for stainless steel as for silver. It also depends on the shape of the implement, but since the business ends of both spoons and knives are essentially thin plates they would not differ much in this respect.

Q's recommendation, were he to bother with such trivia, would be to use a wooden spoon, ideally one of the long thin paddles sometimes provided with takeaway hot drinks. Wood generally has a specific heat capacity of the same order as metal, but a very much lower thermal conductivity.

Softwoods generally have lower thermal conductivities and capacities than hardwoods. Q might also recommend cooling the spoon and wearing gloves. He would certainly advise against employing a gold finger.

John Gee
Aberystwyth, Dyfed, UK

A normal spoon's wider bowl moves more of the martini against the inner surface of the cocktail shaker. Accordingly, heat from the air is conducted more efficiently via the shaker to the drink. Stirring with a spoon is also likely to result in more of the liquid riding high up the inner surface of the glass's cone-shaped bowl.

Still, the case is not closed. Attention to scientific detail demands that the phenomenon be subject to further investigation and any speculation eliminated. You could try repeating with plastic cutlery or wearing heatproof gloves. Good luck.

Len Winokur
Leeds, UK

I can think of at least five ways that stirring a martini with a wide-bowl spoon could result in a higher-temperature drink than using a long, thin bar spoon. First there is the question of simple calorimetry. Assuming the spoon is at room temperature, it will contain a certain amount of heat: a function of its mass, temperature and specific heat capacity, which is quite high for metals compared with plastic or wood. It will contain more heat than a lighter, metal bar spoon.

Second, because it has a larger surface area it will conduct this heat more quickly into the drink and third – for the same reason – more heat will also be conducted from the warm hand via the wide-bowl spoon compared with the bar spoon. The fourth factor is the shear applied by the larger wide-bowl spoon during mixing, which will also add more heat to the system.

Finally, the more turbulent flow caused by stirring with the wide-bowl spoon will tend to draw more warm air into the drink, further increasing the temperature. If you don't have a bar spoon handy when making that martini, try stirring with a pencil, which has less surface area, less mass, a lower specific heat and will, therefore, conduct less heat from the hand.

Mark Wareing
Halkyn, Flintshire, UK

Martinis never die

Your answers on how different spoons used for stirring James Bond's vodka martini affect the drink's temperature made me think of an additional question. Traditionally martinis are stirred using long bar spoons, as your correspondents noted. However, mixologists use the handle end, where there is a disc of metal, rather than the spoon bowl. This makes the martini less dilute. Why?

Pierre Grogan
London, UK

The attentive 007 will have learned that one reason the wider bowl warms the martini mix more is because of the greater amount of air introduced and brought into contact with it. This is because stirring with the spoon bowl moves more liquid around, creating a whirlpool effect. The liquid's level falls in the centre of the cocktail shaker and rises up the side, meaning greater surface contact with the air.

This air will be cooled by the liquid that it comes into contact with, and the more this happens, the more of the air's water vapour will condense into the mix, diluting it.

Also, because alcohol has a lower boiling point than water, it has a slightly higher evaporation rate and any warming effect of the wider bowl will increase this, leading to further dilution. The narrower handle end doesn't make the mix any stronger than it was to start with, but it does minimise subsequent weakening through dilution.

Bond probably wouldn't care as long as his martini was expertly prepared, but perhaps Q would admire his concentration if he did.

Len Winokur
Leeds, West Yorkshire, UK

I am naught but a rude mechanical engineer, but it seems to me that the key words regarding the stirring of martinis with ice to cool them should be 'convection' and 'latent heat'. The material of the stirrer would seem immaterial: its effect on heat conduction is minimal compared with that of the relative velocity of the cocktail liquid moving past the ice.

Perhaps a martini shaker should have an ice-catching spout and the stirrer should be removed before pouring to minimise this effect. Enough ice should be used to ensure the temperature reaches freezing before it all melts.

Latent heat – the heat released or absorbed by a thermodynamic system that occurs without a change in temperature – ensures this is possible. And yes, I have tried it.

David Sherwin
Perth, Western Australia

? Swirled, not stirred

Why shake or stir vodka martinis at all? Both methods mean there is far more collision between liquid and ice, which makes

the martini very dilute. I just hold the cocktail shaker gently and loosely spin it in a circular motion to swirl drink and ice together. Assuming there is a flaw in my method, what is it?

Liam Case
Aylesbury, Buckinghamshire, UK

When you mix ice and alcoholic liquor you want the coldest drink in the shortest time with the least dilution. Professionals start with all the elements pre-chilled: liquor and mixer, cocktail shaker and glass, and with the ice already chipped. The ice, liquor and mixer are added to the cocktail shaker and shaken or stirred vigorously for 15 seconds, decanted through a sieve into the cold glass and given to the customer. Some leave the ice in the drink, which keeps it cool longer, but dilutes the drink as it melts. Choice rules here.

Shaking with chipped ice brings the mix to the coldest point in the fastest time via fully turbulent mixing. It provides a large contact area between the ice and the fluid for fast heat transfer.

Some people use a specialised freezer to make cubes of the alcohol (bottled liquor freezes out water as slush at about –40 °C and reaches full solidification at about –114 °C). As a young student, I did this in the lab freezer – naughty boy that I was.

Bill Jackson
Toronto, Canada

Now you know all there is to know about the vodka martini and its complex science. So mix one, sit back and ponder this final perplexing aspect of the James Bond lifestyle: just how does he stay alive? – Ed.

Die another day

James Bond must have had thousands of bullets aimed at him over the course of his movie career. He must be the most shot-at fictional film hero of all. Can anybody calculate the odds of him not having taken a fatal hit over the past five decades? Surely they are astronomical.

Peter Hammond
Stuttgart, Germany

There is some ambiguity over how many gunshots have been fired at James Bond because in many gunfights it is not clear who the shots are aimed at. However, by my reckoning, in the 22 Bond films to date, there have been at least 4,662 shots fired at our hero. A static well-aimed shot would almost certainly have proved lethal, but assuming all 4,662 were 'on the run', the probability of a single fatal shot is about 5 per cent. That is, the chance of a single shot missing is 0.95, and hence the probability of all shots missing is 0.95^{4662} or 1.4×10^{-104}, which is as close to zero as makes no difference.

Apart from gunshots there have been 130 dastardly attempts to kill Bond. Factor that in, and you have a really small probability. For the record, Bond has also slain 198 villains, creating yet another bizarre improbability.

Gordon Stanger
Solomon Islands

What is even more remarkable about the statistics involved here, is that somebody has bothered to count the shots fired at Bond – Ed.

An earlier correspondent noted that James Bond had survived at least 4,662 gunshots. This figure may not be too far-fetched.

A study of soldiers' small arms hit rates during peacetime found them to be about 60 per cent. This figure would lead one

to expect that the 6,462 rounds fired by British marines at the start of the Falklands war 30 years ago would have produced more Argentinian casualties than the two confirmed deaths and two confirmed wounded. This apparently low figure can be explained by the fact that many shots are fired to keep the enemy's head down and to stop the enemy from firing back.

James Bond, continue the good work. Your chances of surviving to an old age are as good as ever!

Ted Lovesey
Stoke Gabriel, Devon, UK

5 Domestic science

Unmusical scale

My kettle becomes very noisy when it boils. I live in a hard-water area, so I regularly treat it to remove the build-up of limescale. For a few days it then seems much smoother and quieter. What's going on?

Robert Bull
By email, no address supplied

I assume you refer to simmering, the noisy stage of hissing or 'singing' when water explosively forms steam bubbles against the heating element, but the surrounding water is cool enough to implode the bubbles quickly after they form. That cyclic process produces the sound, with the kettle body acting as a soundboard. If you are unsure that this is causing the noise, gently slosh the water in the kettle back and forth when the noise starts. If it really is simmering, sloshing interrupts the noise until the proper, comparatively quiet, boiling gets under way.

In a clean kettle, heat passes efficiently through the heating surface, rapidly completing the simmering stage, so singing soon stops. Also, a clean, smooth heating surface retains bubbles poorly; they detach while small and implode in the water instead of causing large impacts against the heating surface, so the soundboard produces little sound.

However, limescale is a poorer conductor than the heating element or kettle wall, and a thick mass of scale accumulates

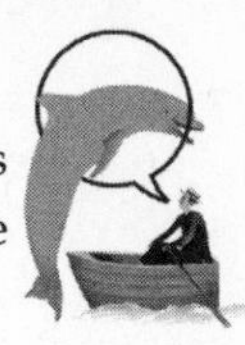

heat before passing it on. Accordingly, the simmering starts slowly, but boiling starts even more slowly, so that the simmering lasts longer.

Jon Richfield
Somerset West, South Africa

Cool bugs

We keep foodstuffs in the fridge to reduce bacterial spoilage but are there any bacteria that thrive best at fridge temperatures? Are some foodstuffs more likely to spoil in the fridge rather than out of it because they carry such bacteria?

Peter Hunt
Teignmouth, Devon, UK

The temperature inside a refrigerator should be kept between 1 and 4 °C – cold enough to slow the metabolism and reproduction of most bacteria while not causing water to freeze, which ruptures cells and damages food.

However, there are so-called psychrotrophic bacteria – including *Listeria monocytogenes* and *Yersinia enterocolitica* – that can reproduce at these temperatures. This led researchers in France to the controversial suggestion that Crohn's disease, in which the gastrointestinal tract becomes inflamed, is on the rise because of the increased use of refrigerators (*The Lancet*, vol. 362, p. 2012).

At 0 °C, even the growth of psychrotrophic bacteria comes to a virtual halt, but microbial activity does not stop altogether. At –2 °C the plant pathogen *Pseudomonas syringae* is still able to make proteins that help water freeze. These can damage fruit and vegetables by causing ice crystals to form near the walls of epithelial cells.

Now consider that bacteria like *Carnobacterium pleistocenium*, *Chryseobacterium greenlandensis* and *Herminiimonas glaciei* have been revived after many thousands of years in suspended animation, locked in ice. The finding has encouraged astrobiologists investigating the possibility of extraterrestrial life. Some believe in panspermia, the idea that bacteria can survive a voyage through space and can seed life wherever conditions are suitable. For these microbes, the inside of your fridge might seem like paradise.

Mike Follows
Willenhall, West Midlands, UK

? Leaf me alone

Why does my dishwasher have such difficulty cleaning spinach from my plates?

Frank Schanze
Stuttgart, Germany

Tinned spinach is used as one of the food residues in standardised tests for dishwasher performance – precisely because it is so difficult to remove. Plus every speck of the dark green vegetable shows up against an otherwise clean white plate. Other tricky foods include mashed potato and coffee grounds, but nothing beats spinach. Every trace left on the plate is a minus point, and these are aggregated to determine overall performance. For several years I was a design engineer on dishwashers, and spinach was our worst enemy.

The tiniest fragment shows up clearly on washed plates. It forms thin flakes which adhere under wet conditions (possibly in the fluid boundary layer, or by chemical bonding, or some other mechanism) and they bake on tightly during

the drying cycle. Even under normal conditions by no means all the crockery surface is scoured by direct jet impact – some is only gently washed with a percolating rain, so there is plenty of space for spinach to lurk.

Furthermore, some of our tests showed that a lot of food residue is redeposited after being removed. This creates other problems, since water volumes during washes are decreasing because of pressure to design machines that use less water (and hence less electricity). At the same time dishwashers are increasing in capacity. You can get more plates in, so more spinach fragments can be deposited than in a previous generation of appliances. Nor will the internal filters trap particles this small. So it just goes round and round the wash cavity.

So what is the solution? I recommend you try a setting that gives you an automatic pre-rinse, or a larger water volume (perhaps not using the 'eco-wash' setting), and definitely use 'rinse aid' liquid if the machine accepts it. Also, try rinsing the plates before you put them in the dishwasher. Studies show that about half the population rinses, and the other doesn't. It is important not to pack the dishwasher: leave every second plate slot open. This can also work well for greasy foods. Try eating fresh spinach instead of tinned. Cook it lightly and the plates will be easier to clean.

Either way, you would be one of the few people I've met who actually eat spinach often enough to notice the effect on wash performance: I was beginning to think that the wash test was totally contrived.

John
By email, no surname or address supplied

Piles of reasons

Why do rugs placed on carpets appear to move of their own volition? How can a bunch of inert fibres exert the force needed to move my rug 15 centimetres in just two weeks when there is a substantial armchair on it?

Ann Duncombe
Falkirk, Stirlingshire, UK

A carpet has a pile direction. The pile does not sit upright but points a little in one direction, giving a slightly smoother feel when stroked with the pile rather than against it. When something like a rug is placed on a carpet and walked upon, the pile bends in that direction, carrying the rug with it. With the release of foot pressure there is no longer the same amount of traction between rug and carpet, so as the pile returns to an upright position it does not drag the rug back with it.

Traditionally, carpets are laid with the pile towards the door. This is from the pre-vacuum cleaner days when carpets were brushed clean and it was easier to brush the dirt out of the door in the direction of the pile.

Max Lang
Northampton, UK

I too have various rugs which travel rapidly on a cut-pile carpet, eventually pushing up against the skirting board.

After a period of careful observation I noticed that the rug always travels in the direction that the pile leans in, and only when someone walks on it. When the foot is removed, the flattened pile returns to its rest position but, like a ratchet, does not move the rug back to its starting point.

The movement is noticeable on cut-pile carpets, but hardly at all on loop-pile carpets, suggesting the effect depends on the pile structure, the length and stiffness of the

fibres, the pressure of the footstep and perhaps the structure of the underside of the rug. Incidentally, it is easier to drag a rug – with or without a chair on top – across a carpet in the direction of the pile rather than against it. I have not calculated the forces involved but they must be fairly large to move an armchair.

I feel an Ig Nobel award approaching.

Donald Brown
Kellas, Angus, UK

We have mentioned the Ig Nobel before. Readers interested in finding out just what an Ig Nobel award is should visit www.improbable.com/ig/ – Ed.

Blank look

My digital TV has a menu containing a guide to several days' worth of forthcoming programmes on each channel. When I first bring up the guide, there are lots of blanks in the schedule. Over the next few minutes these gaps are gradually filled, seemingly at random. How does this happen and what determines the order in which the holes in the schedule are filled?

Alex Andrews
Newton Abbott, Devon, UK

Digital television is sent as a series of packets of information, each containing a block of compressed video, audio or the TV programme data your questioner mentions. These packets are interleaved to produce a stream that can be picked up quickly after selecting the channel.

The amount of compressed data generated by audio and particularly video varies: slowly changing scenes need less bandwidth than fast-changing ones. So the TV programme data is chopped up and transmitted when there is spare capacity left by the other data streams. Since it is effectively transmitted one line at a time, it does not have to be sent in any order.

Alec Cawley
Newbury, Berkshire, UK

The answer depends on the type of box you have to receive television programmes. Free-to-air boxes are probably receive-only devices so cannot request specific data. I suspect the whole schedule is broadcast in repeated chunks and the box has to wait for the relevant data to be received before showing it, similar to waiting for a page to come round in the UK's teletext system.

With cable boxes, the transfer of data from a server to your box takes time. If you had to wait for this information for every page you wanted to skip before reaching your desired page of TV programmes you would quickly get frustrated by how slow it is. To combat this, when you move to a new page a request for that page's data is sent in the background. If you are still viewing that page when the data arrives it is displayed. If, however, you have moved to another page, the data for the previous page is either discarded or cached locally in case you go back. This is a technique called 'lazy loading'.

If you have a cable box and don't see all the programmes on the page appear together then this suggests poor implementation, probably because the code to update the display is inefficient.

For both box types the order in which the information appears on the screen depends on the way the server's software is written. The order that TV programme information is sent in the data packets could be determined by factors such as the order in which it was originally added to the

database, or which data the database server has previously cached to improve performance.

Peter Morris
Birmingham, UK

Rising heat

In hot weather my loft can be almost 20 °C warmer than my cellar, three floors below. Is there anything useful or interesting I can do with this?

David Clarke
Manchester, UK

The best thing you can do with the 20 °C temperature difference between loft and cellar is to make sure you store your food and drink in the cellar and hang your washing up to dry in the attic.

In the grand scheme of things, a 20 °C temperature difference is not very big. If you want to extract mechanical energy from it, then the maximum possible efficiency is that given by a Carnot cycle engine. Running between a hot reservoir of temperature of say 30 °C (303 kelvin) and the cellar at say 10 °C (283 K), then the fraction of the heat flow that can be extracted as useful work can be calculated by dividing 283 by 303 and subtracting the answer from 1. This gives 0.066 or, in this case, 6.6 per cent. This means that for every 100 joules of heat energy that flowed from attic to cellar, you would only be able to extract 6.6 joules as work. In reality it would be even less than this since this is the upper thermodynamic limit for a perfect machine.

This efficiency limit is why practical engines always use high-temperature sources of energy, such as coal combustion.

Running between combustion flames at 800 °C (1173 K) and cooling water at 30 °C (303 K) gives a much greater theoretical efficiency limit of 74 per cent. This is also the reason why using underground hot rocks as an alternative source of energy is a challenge for engineers. The temperatures of the extracted fluids are often only in the range of a few hundred degrees C, so thermal efficiencies are low.

Simon Iveson
Faculty of Engineering and Built Environment
University of Newcastle
New South Wales, Australia

Thermocouples are a proven way of generating energy from a temperature difference between a heat sink and source; they have been used in spacecraft for this purpose and are both practicable and reasonably efficient.

In both cellar and loft, old central heating radiators – the bigger, the better – might be used to feed heat between thermocouples attached to them and their surroundings. Copper and aluminium are potential materials for the connections between loft and cellar. Ideally, several pairs of cables connecting the thermocouples in series would generate a useful voltage for charging batteries in cellphones, cameras or emergency lighting.

Chris Collins
Llandrindod Wells, Powys, UK

Brew beer in the loft, and store the product in the cellar. Job done.

David Everton
Nottingham University, UK

The on-off switch

Occasionally our TV remote control stops working, but a quick jiggle of the batteries usually solves the problem. Why? What impact can this have on how they work or on the circuitry inside the remote control itself? These particular batteries have been in use for more than 12 months, so they may well be nearing the end of their useful lives.

Nia and Steven Faulder
By email, no address supplied

The contacts on the remote handset, which connect to the batteries, are usually made of brass. This means they can oxidise, causing an interruption of the electrical circuit. During jiggling, the batteries will move around, scraping through the oxide layer. This re-establishes contact and the remote control functions again. In more critical equipment oxidisation is avoided by gold-plating the contacts.

Joop van Montfoort
Croyde, Devon, UK

Remotes have microprocessors and, like any computer, they can freeze or lock up. This is usually due to a weird combination of key-presses that the program writer didn't envisage. Removing the batteries reboots the program and normal service is resumed.

As a TV aerial engineer, people's use of remotes always amuses me. Gender differences are especially humorous. Women, if in difficulties with a remote, give the remote a little subconscious push towards the TV as they press. Men usually shake the remote or bang it on a hard object. And both sexes seem to believe that the TV screen picks up the remote commands, even if the handset is for a set-top box.

Rod Buck
Sheffield, South Yorkshire, UK

The chemical reaction in a battery slowly dissolves the casing and produces hydrogen gas at the central electrode, the bubbles of which break the electrical circuit. To prevent this, an oxidising agent is used to convert the hydrogen into water. Too much of the oxidising agent, however, would also cause the battery casing to dissolve. To prevent this, only a small amount of oxidising agent is added, to ensure it runs out before the casing dissolves. Unfortunately, this means hydrogen bubbles build up on the electrode. Shaking can make these fall off, so the circuit can work again.

Chris Evans
Earby, Lancashire, UK

Mega bacteria

Virtually every antibacterial cleaning product I buy claims it will kill 99.9% of bugs. So my question is what type of bacteria or viruses are the 0.1% that these products can't kill? They really must be the sort of thing that you wouldn't want to meet down a dark alley.

By email, no name or address supplied

This figure is derived from the standard tests used to measure the antibacterial effectiveness of a product. The tests include EN1276, a disinfectant and antiseptics suspension test in which known concentrations of bacteria are treated with a product for a given contact time, usually 5 minutes. You may start off with 5 million bugs per millilitre of test solution and, after treatment, end up with 500 bugs per millilitre – 9,999 out of every 10,000 have been killed. This leads to the figure claimed on the product packaging.

For the EN1276 test, four species of bacteria are checked,

two Gram-positive, and two Gram-negative. These give a good representation of what will happen to other bugs. The species used in EN1276 are *Escherichia coli*, *Staphylococcus aureus*, *Enterococcus hirae* and *Pseudomonas aeruginosa*.

Peter Finan
Haworth, West Yorkshire

Wet, wet, dry

Please help settle an argument. In winter we have to dry our washed clothes on our house radiators. However, there are too many items of laundry in one wash for them all to fit on the radiators, so some have to wait their turn on clothes horses at room temperature. My wife says we should put all the small, quickly drying items such as socks, underwear and synthetic sports gear on the radiators first because these will dry much faster than jeans, jumpers and heavy linen items. I argue that the order they are placed and removed from the radiator makes no difference to the total drying time of all the items. My wife says the total drying time is shorter if you dry the small items first and then the heavy items. Who is right?

Jimmy Hardy
Newcastle, UK

Thanks to the many people who answered this question. Sadly the disagreement between our correspondent and his wife persists in those answers, as the following selection demonstrates. Where are the experimentalists when you need them? – Ed.

Let's say we have washed some nylon underwear and cotton T-shirts. We'll assume the nylon dries in 1 hour on the radiator and the T-shirts take 5 hours. (The exact time will depend on the room's temperature and humidity.)

If you put the nylon items on the radiator first, you will be able to put them away in 1 hour, then put the T-shirts on the radiator and get rid of the clothes horse. If you start with the T-shirts on the radiator, the nylon items will have to sit on the clothes horse for 5 hours. So it is a good idea to start with the fast-drying items. It might not be faster but it is certainly more convenient.

To answer the question directly, the total drying time should be the same no matter which clothes you put on the radiator first.

Tristana Simon
Seaford, Victoria, Australia

We can assume that the heavy items such as jeans would take 36 hours to dry at room temperature, while socks take only 18 hours. Let's also assume that items on the radiator dry in one-sixth of the time they need at room temperature, and that we are only washing jeans and socks. If jeans are placed first on the radiator they will dry in 6 hours. The socks will then be dry in another 12 hours at room temperature, or 2 hours on the radiator. The total drying time is thus 8 hours. If socks go on the radiator first then they will be dry in 3 hours, leaving 5.5 hours for the jeans. In this case the total drying time is 8.5 hours.

If you add a third type of garment – say cardigans that take 24 hours to dry at room temperature – then the sequence of jeans, then cardigans, then socks on the radiator yields a total time of 10.5 hours. The reverse sequence would take 11.4 hours.

So it would seem that both the questioner and his wife are incorrect. My solution? Buy a tumble dryer.

Matthew Shepherd
Banbury, Oxfordshire, UK

Drying time is the same, but the amount of time spent on the clothes horse is reduced if you do the quick-drying ones first. But why are you drying clothes on heated surfaces at all? It can damage the plastic bits such as fasteners in modern clothing, nylon fabrics, and more. Buy some fold-away drying racks and wash your clothes in cold water, and you'll notice their colours will stay brighter for longer.

But if you value your marriage, then the answer to the question is that your spouse is right. Give in on the inconsequential things and you'll have a better shot at winning when it really matters.

Liz Zitzow
By email, no address supplied

Black heads

After a while I find the shower heads in my bathrooms become clogged by black flecks of what is obviously some kind of organic material. A similar material accumulates in my cold-water taps if they have not been used for some time, but in this case it is in the form of a black ribbon. I recall from visits made to water-treatment plants in my student days that the passage of water through a filter leads to the build-up of a zoogloea – a translucent jelly-like layer of organic matter. But if something similar to this process is taking place in the shower, why is the material black and exactly what is it?

David Payne
Penarth, Glamorgan, UK

I know this black jelly-like material well because it formed an integral part of my PhD studies, which were focused on the sort of water filters mentioned above. The material is a biofilm: a jelly made up of a layer of bacteria and their extracellular products. These commonly form at the interface between a solid and a flowing liquid.

The black colouration comes from manganese, which is common in groundwater. When water emerges from a spring, any manganese it contains will be in the form of soluble ions. Part of the treatment process is to oxygenate the water with an air cascade, and in some areas chlorine is added to the water to oxidise both iron and manganese. The insoluble manganese oxides created in this process are filtered out before water enters the distribution system.

An alternative to this chemical oxidation is to allow a biofilm of manganese-oxidising bacteria such as *Leptothrix* species to form within a sand filter as the water passes through the filter. This film can build up into a sheath that, in the case of *Leptothrix discophora*, can be up to 20 times the diameter of the bacterial cells.

The ribbon-like nature of the slime shows that the biofilm has formed under conditions of fluid shear. In effect, the film is stretched out by the flow of water. These ribbons are known to biofilm researchers as 'streamers'. Your questioner is seeing a biofilm of *Leptothrix* coloured black by manganese oxide. The water is safe to drink: manganese toxicity is not a problem in municipal water systems, but it can cause staining in laundry and plumbing. These biofilms are not at all dangerous, just messy.

Chris Hope
Lecturer in oral biology
University of Liverpool, UK

Growth potential

My shower gel proclaims: 'New! Stimulates skin flora.' Is there any benefit in this?

Peter Eaton
Porto, Portugal

It sounds like advertising hype to me. Your 'normal' flora don't need any extra nutrition if your skin is in a generally healthy condition. Each distinct zone of healthy skin has its own stable, dominant flora and meddling with it is risky.

The ideal flora for each zone form an even, adherent coating of a particular combination of strains that perform all sorts of different functions, such as tuning your personal and family varieties of body odour. It also crowds out or repels rival strains that might threaten your health. Over large areas of skin your normal beneficial flora form what amounts to a protective non-stick coating.

Growing too vigorously does no good because overcrowding might cause them either to harm your skin or flake off, leaving footholds for alien pathogens. Furthermore, in microbial ecology one of the most important competitive weapons is the denial of food to rivals. Leucocytes in pus, for example, inhibit germs partly by absorbing iron, and therefore denying it to invasive bacteria and fungi. If your shower gel supplies excess nutrients to your skin, that surplus might fuel an alien invasion.

Jon Richfield
Somerset West, South Africa

What a gas

When I turn off a gas ring on my cooker the flame reduces as you might expect, but then it sometimes extinguishes with a small explosive popping sound. Why does it do that?

David Prichard
Bluff Point, Western Australia

When gas is mixed with air it can burn smoothly or it can explode. The gas is fed to the gas ring by a small venturi mechanism, which uses incoming gas to draw in air for combustion. This is adjusted to give a blue cone-shaped flame which indicates complete combustion.

For obvious safety reasons, the flame is not allowed to flow backwards into the cooker. There are two ways to control the flame. The first is by the outward velocity of the gas, which is higher than the speed at which the flame can grow. This is because the gas is travelling so fast at the point it emerges from the very small hole in the ring – the smaller the hole the faster it travels. The second factor blocking the flame is the cold metal of the cooker, which can cool the gas below its ignition temperature. Both of these effects keep the flame outside the gas ring on the cooker.

When the flame is turned off, the gas velocity falls to zero and the flame front advances part of the way back into the system. The cooling effect of the metal is no longer enough to prevent the flame moving inwards. But once the flame front gets inside the gas ring any remaining gas quickly burns off and a small pop is heard. The same pop, on a much larger scale, can destroy a building filled with a similar mixture of gas and air.

When you turn off the cooker, gas and air are in an explosive ratio, albeit, in this case, not one that should cause any damage.

Bill Jackson
Toronto, Ontario, Canada

Ropey soap

What causes or inhabits the black cracks that form on old, infrequently used bars of soap?

Polly Andrews
Great Yarmouth, Norfolk, UK

This will be the reproductive, vegetative part of *Aspergillus niger*, the same mould responsible for the black jelly-like substance that builds up around the bases of taps and washbasin overflows.

Bar soap is essentially the sodium salt of a fatty acid, which, like table salt, absorbs moisture. Even a brand-new bar is hydrated to an extent, being kept so by its impermeable wrapper and subsequent everyday wetting. But if it is exposed to the air for long periods, it dries out and shrinks, causing its surface to crack. Drying mostly afflicts older bars that have worn small, because at this fiddly size they tend to be abandoned by the sink or bath rather than binned.

A. niger is ubiquitous in soil and its spores are readily dispersed in air, where still, indoor conditions allow them to settle. Soap is typically left in rooms where humidity builds up from hot water usage, and any condensation or splashes accumulate in the soap's cracks, these recesses being slowest to dry. Coupled with cosy indoor temperatures, the mould is encouraged to start growing, in the form of vegetative hyphae that penetrate the soap for fatty nutrition and the visible reproductive film above.

A. niger should not be confused with *Stachybotrys atra* (known also as *S. chartarum* or *S. alternans*), which only grows on cellulose substrates and is rare indoors. Though *A. niger* in high doses can cause bad reactions in people, it has many beneficial uses in food production, medicine, as an agricultural indicator of soil micronutrient content, and in evaluating clinical anti-fungal treatments. Indeed, one product of *A. niger*, gluconic acid, is itself used in cleaning agents.

Len Winokur
Leeds, West Yorkshire, UK

Kitchen alchemy

I have an old silver spoon that I keep in a drawer with other cutlery and utensils. I recently took it out and saw that the bottom of the bowl had developed an interesting pattern, presumably caused by a chemical reaction. There were long blackened patches surrounded by an area of rainbow colours, similar to oil on water. What caused this? The spoon was in a drawer with stainless steel cutlery and utensils of other metals. Plastic and wood could also have come into contact with any of these.

Valerie Cumming
Stafford, UK

Silver loves to get together with sulphur, which may come from wood, paper or even from sulphur dioxide in air. The pattern is due to some areas of the spoon being exposed to sulphur more than others.

The rainbow fringes that can be seen around the marks are caused by layers of silver sulphide, with a thickness comparable to the wavelength of light, just as happens with an oil film sitting on water.

John Woodgate
Rayleigh, Essex, UK

Silver easily reacts with sulphur dioxide in the air, making black silver sulphide. This is the common silver tarnish we see. There is a small amount of this sulphur dioxide in the air because of modern internal combustion engines as well as from burning coal and smelting minerals.

Another source is the paper and cardboard with which shelves are often lined. The cheapest way to break down the lignin in wood used to make paper is the 'kraft sulphate' pulping process, which hydrolyses the lignin and frees the fibres. This leaves a sulphur compound in the paper, which turns it brown, although it is often bleached. This residual compound, plus any from the aforementioned sources, produces a low-intensity effluvium of sulphur dioxide that is driven by stray air currents and blocked by other utensils. It will react with any exposed silver, although these stray currents and blockages mean that this circulation impinges on the metal in a varied manner.

This sulphur compound also causes any pulp books made in the past 75 years to become brown and brittle with age, whereas older pulps made mechanically do not degrade in the same manner.

Bill Jackson
Toronto, Ontario, Canada

Here is a guide explaining how to remove such staining from your silver cutlery: bit.ly/GBrSPe – Ed.

Alarming microwave

If we use our microwave oven for longer than about 30 seconds, our car alarm goes off. Why? The car is at least 20 metres away through two walls. The inside of the microwave is a little corroded and the car has a remote central locking/alarm system.

James Joyce
Southampton, Hampshire, UK

Certain car alarms, such as those fitted to Mazda 6, Toyota Rav4 and Mitsubishi Shogun models, transmit a continuous signal at 2.45 gigahertz at powers of up to 500 milliwatts. The microwaves are picked up by sensors inside the vehicle, which detect changes in intensity to signal the presence of intruders. Microwave ovens also operate at 2.45 GHz. While the power radiated within the oven is typically in the range of 600 to 800 watts, the amount radiated outside the appliance will typically be less than a watt. When your oven is in operation, the microwaves reaching your car may be powerful enough to trigger the sensors inside it, which the alarm system interprets as a disturbance within the vehicle.

It is possible to set a car alarm so that the internal signal generator is disabled. You might also want to have your microwave oven serviced in case there is a serious leak of radiation. If your microwave has damaged shielding, the radiated power could be higher than the values above.

Joel Smith
Pateley Bridge, North Yorkshire, UK

It is odd that your microwave is leaking enough radiation to trigger the car's alarm, considering the legal limit – in the US, at least – for leaked radiation from a microwave oven is 1 milliwatt per square centimetre at a distance of 5 centimetres. Perhaps your microwave has a serious leak, or you have an unusually sensitive car. You could try parking the car in front of a friend's house and running their microwave oven to see what happens. If it appears to be solely your problem, consider getting the microwave replaced.

Alex Reinhart
Boerne, Texas, US

6 Our planet, our universe

Moon blues

I was watching the sun set in Crete, as it went through all shades of red in the final minutes. Then I glanced at the moon and saw it had turned blue. How was this possible, and is it the basis of the saying 'once in a blue moon'?

Patrick Casement
London, UK

The chances are that this blue moon was an illusion – the after-image of the sun created by your eyes and brain.

By staring at the bright red disc of the setting sun for a significant amount of time, say more than 20 seconds, the red photoreceptors in your eyes become desensitised. When you switch your gaze to a white object like the moon, the after-image of the sun is seen in its complementary colour – cyan or light blue.

This is because the response of the red photoreceptors has been reduced, leading to the perception that all or part of the red light has been removed; and white light minus red gives cyan. The more yellow that was mixed into the sunset, the bluer the moon would have looked. This illusion is all the more convincing because the discs of the moon and sun we see are the same size so fill the same amount of space in our vision.

Genuine blue moons can appear if volcanic eruptions or fires inject particles with a fairly uniform diameter of around a micrometre into the atmosphere. This diameter is just bigger than the wavelength of red light, which is around 650 nanometres. For example, the particles released by the 1883 eruption of Krakatoa caused the moon to appear blue for nearly two years.

Muskeg, or peat bog, fires have the same effect. Fires that had been smouldering for several years in Alberta, Canada, flared up on 23 September 1950. This produced oily droplets about a micrometre in diameter, which scattered light at the red end of the visible spectrum. With red light scattered out of the line of sight, the discs of both the sun and the moon looked blue, at least when the smoke cleared so that they could be seen.

Of course, it might be tempting to ascribe the blue moon in Crete to the forest fires that plague the Mediterranean.

Mike Follows
Willenhall, West Midlands

You can replicate this effect at home. Stare at a computer screen showing a bright red image for a couple of minutes and then immediately glance at a white sheet of paper – it will appear to be bright blue for 30 seconds or so before your eyes readjust.

Mike Sparks
By email, no address supplied

The term 'blue moon' comes from the traditional agricultural naming of the full moons throughout the year.

The 12 full moons we see each year are named according to their relationship with the equinoxes and solstices. The names vary in different regions, but well-known examples are the harvest moon, which is the first full moon after the autumnal equinox, and the hunter's moon, which is the second full moon after the autumnal equinox. Similarly the Lenten moon,

the last full moon of winter, is always in Lent, and the egg moon (or the Easter moon, or paschal moon), which is the first full moon of spring, is always in the week before Easter.

By this system there are usually three full moons between an equinox and a solstice, or vice versa. However, because the lunar cycle is slightly too short for there to always be three full moons in this stretch of time, occasionally there are four full moons. When this happens, to ensure that the full moons continue to be named correctly with respect to the solstices and equinoxes, the third of the four full moons is called a blue moon.

There are seven blue moons in every 19-year period. The last blue moon was on 21 November 2010, and the next will be on 21 August 2013.

Aidan Copeland
Chester, UK

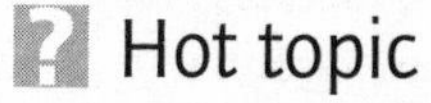

Hot topic

One suggestion to combat climate change is that we should become vegetarians as livestock is more environmentally damaging than growing crops. However, if we stopped eating meat, livestock would still live, so is the suggestion correct? Or are we expected to cull any remaining pigs and cows?

Ella Gribben
London, UK

World population now stands at almost 7 billion, and is projected to level off at around 10 billion by 2050. The world currently has 17.3 million square kilometres of cultivated land; that area of land could feed 3 billion people eating mainly meat, or 42 billion people eating a strict vegan diet of mainly potatoes. For a more varied vegetarian diet, with vegetables, fruit, eggs and milk, we could feed up to 12 billion people on current world farmland and we could also eat seafood. On the negative side, these figures make no allowance for food waste or losses from disease or other disruptions to the food chain.

Farm animals produce much else besides meat, such as wool, leather, milk, cheese and eggs; so even in a vegetarian world, some animals would be retained, but far fewer. The lifespan of most farm animals is quite short, between 10 and 20 years at the most and often fewer. So without breeding, their population would fall quickly. Farm animals produce carbon dioxide but only as much as was locked up in the grass they ate. They also produce methane, another greenhouse gas, but so do rice paddies and submerged vegetation in reservoirs.

The main benefits of a vegetarian diet with respect to global warming would be a reduction in emissions from the energy used to transport meat from farm to fork: because less land is needed to grow vegetables, more food could be produced locally. However, intensive greenhousing, such as in the Netherlands, can be very energy intensive, especially as consumers insist on having many vegetables available on supermarket shelves all year round, even out of season.

The main environmental benefit from everyone going vegetarian would not be reduced global warming but less pressure on world ecosystems – terrestrial and marine – to feed 10 billion mouths. Indirectly, global warming might be reduced as less rainforest is felled, less energy is devoted to food transport and less energy is needed for intensive farming methods. On the 'downside', we might all be healthier and less obese, so we'd live longer and world population would be increased somewhat.

Hillary Shaw
Senior lecturer and food research consultant
Harper Adams University College
Newport, Shropshire, UK

Few livestock are kept for longer than two years before slaughter for meat, and because we farmers respond readily to the economics of supply and demand, it wouldn't be long before decreasing consumer demand for meat was reflected in smaller numbers of farm animals.

The questioner's statement that 'livestock is more environmentally damaging than growing crops' begs many questions. A grass-fed suckler operation for beef, for example, produces less carbon than a similarly sized arable farm. Livestock can be integrated into vegetable-growing systems in order to reduce the amount of climate-damaging tillage and diesel use.

To farm sustainably and minimise machine tillage, farms will probably still need livestock, albeit in far smaller numbers.

Chris Smaje
Frome, Somerset, UK

The largest proportion of any crop is inedible for humans. All that biomass has to go somewhere, and the easy thing to do is feed it to animals that process it into meat. The sheer volume of low-grade waste is too great to process for biofuel and the cost for other forms of disposal would be massive. If it is not removed, the next crop cycle can't be planted. Rotting plants produce greenhouse gases too.

Other animals will move in, multiply hugely and deal with the biomass. They will also produce the same greenhouse gases as cattle do now. In Australia this will probably fall to rabbits, kangaroos, camels, wild pigs and buffalo.

As my university lecturer says: 'Given the same area of land, you can feed a lot more people on bread and pea soup than you can on steak, but you can do it for a lot longer and with fewer problems if everyone gets bread, pea soup... and some steak.' The cattle eat the pea vines and live on the pasture, which is used to rotate the wheat and pea plots. The dung and urine can fertilise the soil and humans get to eat a steak here and there.

This is a complex process without an easy solution. So don't feel guilty eating the odd steak. Just ensure it was grown largely on pasture and crop residues and not transported too far.

Jan Horton
West Launceston, Tasmania, Australia

Living in the past

If a camera was placed 1 light year away from Earth with a high enough definition, could it be used to spy on events that took place on Earth one year ago? And, if so, could this technique be used to record our past by sending an array of such cameras to the appropriate distance in order to capture momentous events in Earth's history?

Jon King
Swindon, Wiltshire, UK

In theory, yes, a camera placed 1 light year away could indeed record events on Earth a year earlier. Of course, because it would need to transmit its information back to Earth via radio waves, which also travel at light speed, it would take another year for the images to be returned.

The real problem would be positioning the camera at that distance. The fastest you could move the camera to that position would be at some fraction of the speed of light. Even if you could send your camera out at light speed, the camera will never 'catch up' to light that left Earth before the camera did, so the camera can only capture images of events

which happen after it is launched. In other words, it amounts to an elaborate and expensive mechanism for recording the present, but in a way that means you can't view the results until a year later. You could record the present much more easily with an earthbound camera, and wait a year to view it.

Agim Perolli
Carmel, New York, US

Although we can't outpace and photograph the light that left Earth at any time in the past, if we located a black hole, say 50 light years away, and if its surroundings were perfectly free of obscuring gas and dust, then in principle we could see light from Earth bent around the hole. This is because black holes have the gravitational strength to bend light through 180 degrees – and direct it back towards us, giving us a view of events 100 years in the past.

Achieving this might be a bit challenging, though.

Greg Egan
Perth, Western Australia

Building a camera with the necessary resolution would be no mean feat. Creating a sensor with enough megapixels, or a roll of film with fine enough grain would be hard, but the trickiest part would be making the lens. The maximum resolution a lens can attain is limited by its size. To recognise objects a centimetre across, from a light year away, you would need a lens several hundred times the size of the solar system.

To give you an idea of how big a limitation this is, take a look at the best pictures we have of Pluto: bit.ly/hVhVDd.

James Pickering
Gateshead, Tyne and Wear, UK

Yellow to the core

Why do we perceive the sun as yellow? I understand that its colour arises from the mixture of wavelengths in sunlight. But as sunlight has been the main source of illumination throughout evolution – the background light to everything on the planet up to the invention of electric light – it gives us our default colour for everything. So why don't we consider it to be the neutral colour, in other words, white?

Perry Bebbington
Kimberley, Nottinghamshire, UK

Why do we perceive the sun as yellow? Because the sky is blue.

The sun's light is white, but when it hits the Earth's atmosphere it is scattered by the atoms, molecules and dust in the air. If this didn't happen the sky would be black, as it is in space. This was first explained by John Strutt, who became Lord Rayleigh, which is why the process is known as Rayleigh scattering.

Shorter wavelengths, such as violet and blue, are scattered more than longer ones. Because blue light is scattered across the entire sky – turning it blue – the remaining direct light from the sun appears richer in longer wavelengths such as yellow and red. When the sun is very low in the sky its light passes through more of the atmosphere, resulting in more of the spectrum being lost to scattering, making it appear more orange or red.

On a cloudy day we see everything in its true colours because there is no direct light and all wavelengths are transmitted equally. Our eyes automatically adapt to the colour balance around them so we don't notice this, but photographic colour film can't do that. It is adjusted for sunshine, so cloudy photos will turn out blue.

Our eyes can even adapt to tungsten artificial light, but to capture this in a photo we need to use a film balanced for artificial light. Of course, as people switch to digital photography, it is much easier to correct for these problems.

Guy Cox
Centre for Microscopy & Microanalysis
University of Sydney
New South Wales, Australia

The writer is the author of Optical Imaging Techniques in Cell Biology, *published by CRC Press/Taylor & Francis (2006) – Ed.*

I contest the statement that we see the sun as yellow. Most people never look directly at the sun – it is dangerous and virtually impossible to do so. The only time we look at it directly is when it is near the horizon, where it does appear to be a variety of shades from yellow to red thanks to the scattering of blue light. But at those times that part of the world takes on the same orangey hues, as that is the colour of the source.

The definition of a white surface is one that reflects all wavelengths of the visible spectrum equally. The human eyes and brain see such a surface as white under a wide range of conditions, such as under standard incandescent light bulbs. This is true even though such a light is in fact yellow, as one can observe by seeing a lit room from outdoors at dusk. But we still see the surface as white.

The same surface viewed under the noon sun will still appear white, but it is not the same white as that seen under the artificial light. Most importantly, it will not appear yellow, showing that the sun itself is not yellow – rather, it actually defines the whitest white that we ever see. So I would argue that the sun is white (but please don't look at it to check).

John Elliott
Stockport, Cheshire, UK

Sound of silence

My daughter dived underwater in the swimming pool and screamed as loud as she could. I was right next to her with my head out of the water, but I could only detect the tiniest sound at the end of the scream. But when I was underwater with her, I could hear most of the scream. Why?

Linda Simpson
Katonah, New York, US

This is an example of impedance mismatch. Acoustic impedance – a measure of the way a sound wave interacts with the medium it is passing through – varies depending on the medium concerned. When a wave encounters a medium with a different impedance from the one it is in, most of its energy will be reflected at the boundary.

Most people will be familiar with ultrasound scans of a developing fetus. For these scans, a gel is applied to the expectant mother's skin to reduce the impedance mismatch between her body and the transducer and so maximise the transfer of acoustic energy.

Despite the acoustic mismatch between air and water, our everyday experience is that sound passing through the air manages to reach receptors immersed in the fluid of the cochlea within the inner ear. For this, we can thank the design of the middle ear for impedance-matching the sound.

Vibrations are passed from the eardrum via the auditory ossicles – the three smallest bones in the human body – to a membrane in front of the cochlea called the oval window. The eardrum is attached to the first of the ossicles, called the malleus (or hammer). This is pivoted about the second bone, the incus (or anvil), which in turn is fused to the third bone, the stapes (or stirrup). The stapes drums on the oval window.

The eardrum has an area about 15 times that of the

oval window. As the sound energy striking the eardrum is concentrated at the oval window, the amplitude of the sound vibrations is increased. Lever action further boosts this amplification. Without the middle ear ossicles, barely 0.1 per cent of the energy arriving at the eardrum would reach the inner ear.

Mike Follows
Willenhall, West Midlands, UK

Flash flood

I was visiting Shropshire, in England, when I woke abruptly in the early morning. Outside it was hot and humid. As I was sitting facing the window I saw a bright single-point flash of vivid blue light through the thin curtains, just as if I had looked at a photographer's flash going off. Then there was a single heavy rumble of thunder followed by a sudden and intense downpour of rain, lasting only a few minutes. Then all was quiet. Later, looking through the same window, I realised the location of the flash was a lightning conductor on the roof of a school. Talking to others I discovered the whole event had been very localised, not even extending across the village. I thought thunderstorms worked on a larger scale. How had there just been the single flash at the lightning conductor and such a small-scale storm, contained within a hundred metres of the school?

Stephen Huyshe-Shires
Sidmouth, Devon, UK

In many areas in Africa one may sit on a hilltop overlooking expanses of veldt enjoying a panorama of well-separated thunderheads, each occupying a cell a few hundred metres across, precipitating bolts and showers on all beneath.

In South Africa's drought-prone Namaqualand they call

such a cell 'jaloersreën', which translates from the Afrikaans as 'envy rain' – each farmer thinks the shower is over his neighbour's property and envies the other.

This type of thunderstorm often occurs when conditions favour strong, localised updraughts without much lateral wind. However, on its own such an updraught might bring no rain. It is heavy ionisation from a lightning bolt that is to blame. This disrupts the uniform electric charges on droplets in the thundercloud, causing many of them to attract other droplets with different charges. When droplets grow large enough their aerodynamic behaviour changes and they collide with smaller neighbours, growing increasingly bigger until there is a bout of heavy rain. If the updraught is strong enough, it might cause a hailstorm.

The Shropshire school storm sounds like an unusually small cell of such a type, triggered by the lightning conductor.

Jon Richfield
Somerset West, South Africa

7 Feeling OK?

Shout it out

Why do I become hoarse if I shout? And how does my voice recover?

Bryan Patrick
Aberdeen, UK

Shouting injures the vocal cords mechanically as it requires a large volume of air to pass at high velocity through the windpipe from the lungs. The injury is in the form of inflammation and swelling of the mucous membrane of the vocal cords. This causes the voice to become hoarse. Vocal rest for a few days will allow recovery as the inflammation subsides.

Kameshawar Rao Matcha
London, UK

People with trained voices, such as singers and actors, can often produce amazing volumes of sound without much effort and with very little harm or discomfort to their throats.

Roughly speaking they form their mouths and throat from the larynx upwards into an exponential horn, the most efficient shape for sound production. Suitably coupled with the vibration of air, a small amount of vocal energy can produce a formidable volume.

The untrained voice produces louder sounds by brute force rather than technique. We force air more violently between our vocal cords, thereby damaging them. Cords react largely by coating themselves with more and gummier protective mucus, and by fluid swelling. Both effects interfere with their correct, efficient vibration and stop them closing and opening in the proper way, causing hoarseness.

If you are lucky enough to escape bacterial infection in the damaged tissues, then a little rest – perhaps a day or two – will give the tissues time to recover. Should you persist in such abuse, you risk permanent injury, though your vocal cords will very likely become calloused and you will stop becoming hoarse.

However, if you are in a profession that demands continual shouting, I strongly recommend taking a course of voice training. Apart from escaping injury, you will be able to make yourself heard with considerably less effort.

Jon Richfield
Somerset West, South Africa

Split times

Why does cold weather dry my skin out? Generally things dry out more slowly the colder it is, yet as soon as winter arrives my hands become so dry the skin splits.

Bernard Marie
Lille, France

Air at a given temperature can hold only a certain amount of water vapour: the colder it is, the less it can hold. In cold weather, this saturation level is so low there is very little water vapour in the air, even if the relative humidity (the ratio of the amount in the air to the saturation value) is high.

The rate of evaporation from a moist surface depends on the difference between the concentration of vapour in the

air right at the surface and the concentration in the bulk of the air. The former depends on the temperature of the moist surface: in the case of your skin, it is warm even in cold weather, whereas the vapour concentration in the bulk is low in cold weather. So evaporation takes place at a higher rate than in warm weather.

This is true even in a heated building, because heating the air that has come in from outside does not increase its water vapour content. Humidifiers can add water, but they consume a lot of energy. Large buildings sometimes have systems that can transfer water between incoming and outgoing air, which are more energy-efficient.

Though water evaporates faster from your skin in cold weather, your laundry dries more slowly. Because it is not heated by your body as your skin is, the level of water vapour right at the surface is quite low, so the rate of evaporation is low too. If you want your laundry to dry quickly, wear it.

Eric Kvaalen
La Courneuve, France

Physician, heal thyself

Do medical doctors, because of their knowledge of medicine and proximity to medical services, have a longer life expectancy than other people?

David Ashboren
By email, no address supplied

Doctors are as interested in their own mortality as any other group of people, and have investigated this very subject.

The authors of a 1997 article in the journal *Occupational and Environmental Medicine* looked at the cause of death listed in UK Department of Health records for more than 20,000 hospital consultants who died between 1962 and 1992. Overall, these doctors had rates of lung cancer, heart disease and diabetes that were less than half those for the general population. The study also found distinct statistical variations depending on the consultant's area of specialisation. For instance, suicide was more common among anaesthetists than other doctors.

While direct access to healthcare may have played a part in doctors' relatively low rates of disease, we are obviously aware of the risks of smoking and an unhealthy lifestyle and, more importantly, have acted on them far more than other people. Would that we could persuade everyone else.

Dr John Davies
Lancaster, UK

According to an American study covering the period from 1995 to 1999, doctors do indeed live longer than the average for their corresponding national population, but whether this is because of their knowledge of medicine and proximity to medical services is another matter. After all, it is well known that socio-economic status and educational attainment have a big impact on health and hence life expectancy. A study of the UK population showed that middle-class professionals such as doctors and accountants outlive builders and cleaners by eight years on average. Another American study suggests that doctors live longer than other professionals, but there is no clear evidence yet as to why this should be the case.

Of course, we doctors do use our medical knowledge to try to understand any issues affecting our own bodies and minds, and we may then try to treat ourselves or seek help from our colleagues. But we are not, in fact, good at seeking help, being patients or taking time off work when we are sick. Recent statistics on the UK's National Health Service workforce show that doctors are much less likely than other healthcare professionals to be off work, but don't assume that is because we are less likely to be sick. As a doctor who has turned up for work despite feeling poorly, I know what I'm talking about.

Dr Joanna Jastrzebska
North Shields, Tyne & Wear, UK

Pine flu

People can be infected by bacteria, viruses, fungi and animal parasites, but are any human or animal diseases caused by plants? Is it possible to suffer a moss infection, come down with a bad case of the ferns, or contract wisteria? If not, why have no plants taken advantage of us in this way?

Bevan Tattersfield
London, UK

Plants do cause disease. Think of allergies to plants such as poison ivy or products such as peanuts. *Ailanthus* (also known as the tree of heaven) bears flowers whose smell has been associated with headaches and nausea.

But these are not infectious diseases. Plants do not become pathogens inside us as they need light for photosynthesis, which they can't get inside our bodies.

Eric Kvaalen
La Courneuve, France

I know of no vascular plants that are truly infectious in animals, though some do infect other plants. H. G. Wells wrote a short story entitled 'The Flowering of the Strange Orchid' in 1894, but that was really about plants preying on humans.

Hairs on some fern species may be carcinogenic if eaten regularly, while pollens may cause hay fever. Cereal crops may host disease-causing fungi such as ergot and aspergillus, and various plants host and transmit pathogenic invertebrates, such as flukes.

Certain barbed grass seeds can catch in animals' hair and sometimes actually burrow through skin. I have seen one that had germinated in the kidney of a little girl. However, the seeds of most plants are too large to be effective as infectious agents; spores might do better. It's also conceivable that evolution might eventually lead to parasitic characteristics in dust-like seeds such as those of witchweed or orchids, or in pollen.

Still, I don't think that these examples capture the spirit of the question.

Jon Richfield
Somerset West, South Africa

There is a case, widely blogged about, of a man with a small fir tree growing in his lungs (bit.ly/1NYo2o). He was cutting pine trees when he inhaled the seed of a cone, which settled in his lung. Soon his body started to form a cyst around the sapling. Doctors feared it was cancerous so opted to remove it, whereupon they found the sapling.

Christopher Payne
Essex, UK

My late uncle, an English country doctor, used to tell the story of a patient with excruciating pain in his face. The culprit turned out to be a tomato seed lodged in a crevice in the roof of his mouth, which had germinated and was growing into his hard palate.

Andrew Cooper
Walls, Shetland, UK

True plants may not have evolved into infectious disease agents, but one of humanity's greatest scourges, the malaria parasite, evolved from a kind of single-celled alga.

The plasmodium parasite that causes malaria has a structure in its cells, called the apicoplast, that is a remnant of a chloroplast once used for photosynthesising. It's not understood what it does, but the parasite cannot survive without it.

Plasmodium is just one member of a large group, the Apicomplexa, whose ancestor gave up photosynthesis and turned to parasitism more than half a billion years ago. They are responsible for a wide range of diseases. Other examples include *Toxoplasma gondii*, which infects a third of the world's population and has been linked to a higher chance of accidents, and *Cryptosporidium*.

Michael Le Page
London, UK

Urine and out

Does cranberry juice cure cystitis? If it does, how does it work – surely urine is just urine by the time it is excreted from the body? If it doesn't, how did the myth arise?

Pauline Greenshiels
Liverpool, UK

Cranberry juice can help cure cystitis thanks to its proanthocyanidins, which have an anti-adhesive effect on bacteria in the bladder.

We have known since the 1980s that bacteria can stick to the bladder wall and bury themselves in mucus – which protects them from antibiotics. So while the antibiotics taken for an infection clear the bladder of free-floating bacteria, the ones hidden in mucus can then come out of the woodwork a fortnight or so after the course of antibiotics is finished. Then the misery of cystitis is back.

Taking cranberry juice can help to prevent this. Many people with cystitis find that cranberry juice on its own works well enough, though not all and some will still need an antibiotic too.

James Wakely
Part-time GP
Colchester, Essex, UK

The cystitis for which cranberry juice is effective is caused by a bacterial infection, most frequently by the gut bacterium *Escherischia coli* but *Staphylococcus saprophyticus* is behind about 1 in 10 cases.

Cranberry juice contains proanthocyanidins that block the ability of these bacteria to adhere to the bladder and urinary tract. The proanthocyanidins alter the molecular structure of the fine protein filaments, or fimbriae, by which the pathogens attach.

As with all remedies, several factors can affect the juice's efficacy, meaning it can vary between individuals and between episodes within one individual. Also, the onset of cranberry-juice intake may simply coincide with the body's natural recovery, while the placebo effect could be involved too. People disposed to drinking cranberry juice may have a healthier diet and lifestyle. Quantity and frequency of intake, and level of infection, will be important and the juice may augment or diminish the action of other medications and vice versa.

If you enjoy drinking cranberry juice as part of a balanced diet, by all means continue. Just don't rely on it as a cure. And certainly not to the exclusion of any medicines you have been prescribed.

Len Winokur
Leeds, West Yorkshire, UK

First, almost everything about the cranberry 'cure' is beset with doubt and controversy. The consensus is that, to the extent that it does work, it does not so much cure cystitis as assist in discouraging recurrence, allegedly by preventing bacteria from adhering to the urinary mucous membrane.

This is not implausible, but it is unclear how much cranberry product is necessary for a useful effect; a couple of dozen glasses of pure cranberry juice per day are proposed, while juice cocktails are considered worthless. That hardly sounds practical. Commercially available capsules seem to be equivalent to a small number of cranberries, so generous helpings on your breakfast cereal should be more economical and perhaps more effective.

Urine is urine? Undeniably so, but urine does vary drastically with diet and physiology. Eating liver turns it yellow; beetroot turns it red; asparagus, cooked mutton, corned beef, coffee and so on, all have olfactory effects, and practically everything affects urinary pH. Urine is the product of your kidneys' action on whatever your body absorbs into the blood, and its character varies accordingly, sometimes with dramatic effects on the normal (or abnormal) microflora of your urinary tract.

Jon Richfield
Somerset West, South Africa

Meta-analysis suggests that cranberry juice may have value for women who have recurrent urinary tract infections. The evidence is not conclusive and further stringently designed studies are needed. To find out more, see the Cochrane Database Systematic Reviews at 1.usa.gov/d525a5 – Ed.

Painless solution

I cut the inside of my finger on a piece of smooth metal foil, but felt nothing. The first I realised I had done it was when I started leaving patches of blood on things. The cut was quite deep and took a while to staunch, so why was it completely painless? I've had similar cuts on sharp objects before and mostly they've been very painful, but every so often some cuts don't seem to hurt at all.

Don Pemberton
Perpignan, France

I'm an anaesthetist, and I 'cut' my patients' hands every day when I insert an intravenous cannula, usually into the back of their hand or wrist. Patients expect it to hurt, as do I, so I warn them. Sometimes afterwards the patient will ask if I have done it yet, and exclaim that I must be good, as it didn't hurt a bit.

I accept this praise modestly, because I know skin sensation is perceived via discrete receptors with free nerve endings used for detecting pain. If I choose an area that happens not to contain pain receptors, the patient may feel the needle pressure but no pain.

John Davies
Consultant anaesthetist
Lancaster, UK

It depends on how cleanly the nerves are severed. For example, the edges of the tins and lids left by a tin opener are rougher than those of metal foil, and tear the nerve endings. It is this tearing of the pain receptors that also makes paper cuts so sore, because the edges of even the shiniest paper are rough at the microscopic scale. Any contamination of the wound with acidic or salty liquid from the tin contents, or with microscopic debris lodged in the offending edge or on the skin, will further irritate the nerves.

Another factor is the number and density of pain receptors on the part of the skin that is cut. Fingers have a very high density of nerve endings, including those for touch and temperature, but even across a given finger tip these are not uniformly distributed. An otherwise identical cut will be more painful if it occurs at a spot where they are more densely packed and more endings get damaged.

There is a circadian – or daily – rhythm in pain threshold. Generally, the least pain is experienced in early to mid-afternoon, with more experienced in the morning and at bedtime. Individuals differ in their tolerance of pain and there is evidence that people adopt similar coping mechanisms to their parents.

Other factors can also influence someone's perceived pain at a given time, through complex interactions between chemical neurotransmitters, hormones and the way in which nerves are hard-wired. Inspecting or applying pressure to the injury, good health and physical fitness, and being immersed in an activity help reduce pain. Negative moods and sleep deprivation, in contrast, tend to worsen it.

About a year ago I had carpal tunnel surgery on my hand, which involved dividing and parting the overlying tissues to access the carpal ligament where the palm joins the wrist. Having expected discomfort as the anaesthetic wore off, I was surprised by the absence of pain. The procedure was carried out by a top surgeon using finest-grade instruments. I was even advised to return to my passion for piano quickly to maintain motility of the tendons, and I know at least one person who, after carpal tunnel release on both hands, was back at work the next day.

Len Winokur
Leeds, West Yorkshire, UK

Achoo blues

Twice, when donating blood platelets to the Red Cross, I've started sneezing uncontrollably. Assistants respond immediately by giving me an antacid indigestion tablet. Their explanation is that this restores my calcium balance. I understand how my calcium could get out of whack when my blood is being removed, filtered and given back to me without platelets, but what does this have to do with sneezing?

Heather Hallen
East Lansing, Michigan, US

Donating platelets involves the use of a plateletpheresis machine, which draws and separates platelets from a donor's blood before returning red cells and plasma to the donor's bloodstream.

A solution called acid citrate dextrose is added to the blood to stop clotting during this process: citrate, in particular, has the effect of lowering blood calcium levels to a point where the blood cannot clot in the machine.

This calcium deficiency, or hypocalcaemia, may cause side effects when blood from the machine is returned to the donor's body. Symptoms include tingling, chills, slight nausea, bruising, fatigue and dizziness; the most frequent one is tingling around the lips.

Sneezing is not a common side effect, but low calcium levels could cause tingling in the nasal passages and hence sneezing.

The treatment for this is first to slow down the rate of platelet collection and then to supply the donor with antacid tablets, a rich source of supplementary calcium. If this is not effective, the procedure may have to be halted until symptoms resolve and a decision made over whether to proceed at a lower rate of platelet collection.

Richard Benjamin
Chief medical officer, American Red Cross
Washington DC, US

One bump or two?

If I hit my head on a blunt object it invariably produces a large lump within a few minutes, yet when I hit, say, my thigh or my hand with the same force, if any swelling occurs at all it is very minor. So why the difference?

Paul Stretton-Dawn
Cambridge, UK

The lump you feel is a bruise. Bruising results from the force of the impact damaging tiny blood vessels or capillaries. Blood escapes into the surrounding tissues to produce swelling. It is this blood that is responsible for the purple colour of visible bruises such as a black eye.

Blows to the head produce such a lump because there is only a thin layer of soft tissue between the skin and skull to cushion the blow. Because the thinner tissues of the scalp dissipate less of the blow, more of the energy translates to capillary damage, leading to profuse bruising.

The thigh, in contrast, has much thicker deformable muscle between the skin and the bone so does not bruise so easily. Even if you hit your thigh harder to produce bruising comparable to that on your scalp, the swelling on your thigh would be less. This is because on your thigh, the leaking blood can perfuse more or less in all directions. Conversely, on your head, the much thinner soft tissue restricts sideways perfusion. This, together with the non-deformable skull, means that tissue distension is predominantly outward, making the lump bigger, with the greater pressure beneath the skin making it feel firmer and more painful.

Also, because the thigh bone is essentially a rod, an oblique blow to the thigh may transfer the shock pattern so that it dissipates almost entirely through the softer fat and muscle, reducing the size of the bruise.

Inadvertent knocks to the hands typically happen with the arm swinging freely, so the energy of the impact translates mainly to movement of the hand or arm about the joint rather than tissue damage. Deliberate raps on a solid surface with even the soft parts of the hand, however, are quite a different story – as anyone who has tried attracting the attention of a neighbour whose front door lacks a knocker will know.

It is for the same reason that blows to the head with a blunt object can break the skin. Hair provides little protection. Similar considerations account for the pain frequently accompanying skin damage after knocking one's shin or stubbing a toe.

Len Winokur
Leeds, West Yorkshire, UK

? Happiness syndrome

It seems that people who describe themselves as happy are less likely to catch a cold than those who say they are unhappy. Even when happy people succumb, they have fewer symptoms. What is going on?

Jack Read
Brisbane, Australia

First let's agree that when we talk about 'happiness' we are not referring to transient, hedonistic pleasure, but a general feeling of well-being and satisfaction with life. This varies between individuals and is what relates in some way to a propensity to catch colds and indeed other illnesses.

We know there is a correlation, so broadly speaking there are three possibilities: being happy makes you more healthy; being well (more often) contributes to feeling happier; other factors affect mood and overall health.

A study of people after flu vaccinations showed happier folk generated more antibodies. Another study showed that the smiles on photographs of novice nuns were good predictors of their longevity – happier ones living longer. Both of these suggest being happy makes you healthier. We could put forward a mechanism by which happy people socialise more, are exposed to a wider range of pathogens (residing in other people) and so strengthen their immune systems.

Recently reported in *New Scientist* is a link between infection and mood or depression. This suggests happiness could be the result of not being sick rather than a cause.

The Positive Psychology movement, spearheaded by Martin Seligman, who made his name by studying depression, makes the economic case for paying attention to well-being. After all, if increasing happiness means fewer illnesses, less time off work, less pressure on medical resources and so on, that has to be good. And if it's the other way round – no harm done.

Pauline Grant
Business psychologist
Beaconsfield, Buckinghamshire, UK

Golden wonder

Almost daily people comment on my deep tan, but I'm a Caucasian Scot who seldom sees the sun and wouldn't get on a sunbed if you paid me. I live in a modern flat lit by dozens of halogen bulbs. There are 36 in the living room alone. Could they be tanning my skin?

Chris Smith
Edinburgh, UK

Possibly. With 36 lamps, the room must be brighter than most, and halogen lamps do produce noticeable amounts of ultraviolet radiation. It's UV from the sun that usually tans human skin. The capsule of a halogen lamp is made of quartz to withstand high temperature and pressure, but quartz transmits ultraviolet, so in domestic lighting cerium is added to absorb this, and a layer of glass provides additional screening. Common halogen lamps are either enclosed in an outer glass envelope, or should have a glass shield in the fitting. Your correspondent could check to see if the shield is present.

Even so, shielding is not 100 per cent effective. Assuming a bright indoor lighting level of 1000 lux, the UV radiation from normal halogen lighting is around 0.1 watts per square metre. At this rate, the erythemal dose, or the level of ultraviolet that causes redness of the skin, could be reached on average in a million seconds, or two continuous weeks. The permissible exposure time, to which nearly all individuals may be repeatedly exposed without adverse effects, is somewhat less.

Sensitivity to UV varies by a factor of 10, so tanning cannot be ruled out. To test for the effect, you can buy ultraviolet radiation detector stickers, but it would be simpler to apply a high-factor sunscreen to a test patch of skin. It is also worth noting that antibiotics, diuretics and coal-tar shampoos increase photosensitivity.

Slight tanning may not be a bad thing, particularly in a Scottish winter, because it boosts vitamin D levels, but if your correspondent is concerned, he could read the following paper and then eat lots of pizzas: 'Tomato paste rich in lycopene protects against cutaneous photodamage in humans in vivo: a randomised controlled trial' (*British Journal of Dermatology*, vol. 164, p. 154).

David Craig
Edinburgh, UK

Whether the previous correspondent is correct or not, it might also be worth paying a visit to your doctor – Ed.

Your correspondent might want to find out if he has haemochromatosis – a hereditary disease caused by the body absorbing too much iron from the diet. It is also known as bronze anaemia because of the tan it can impart and has been dubbed the 'Celtic curse' because it is so common among Scots and Irish.

I too got loads of compliments on my tan until I was diagnosed with the condition and had treatment, which involves venesection, or drawing off blood. Now I'm back to my pale but healthy self. In retrospect, the lack of tan lines around my neck and wrists should have given me a clue.

Kieran Crehan
Dublin, Ireland

A wee dram

We often hear about people surviving where water is scarce by drinking their own urine. But can drinking urine – one's own or anybody else's – harm you? If so, how? Does it have to be fresh? And how many times can a person recycle it? Obviously, it can't last forever.

Peter O'Brien
Dublin, Ireland

Somebody else's urine may not kill you, but it won't do you much good if it is carrying pathogens, pharmaceuticals or food allergens. If you must drink urine, your own is the safer bet. Urine as produced by the kidneys is non-toxic and, urinary tract bacteria aside, your own will be free of anything alien.

The main solute in urine is urea, a harmless neutral end-product of protein metabolism and the ammonia this generates. The characteristic yellow colour is down to urobilin, the breakdown product of spent red blood cells. The urea will start oxidising back to ammonia on exposure to the air though, producing the smell characteristic of urinals.

One thing your body definitely won't want back are the salts that urine contains, which is why drinking urine is a bad survival tactic. It is not just the decreasing volume and increasing concentration with successive passes that is of concern. Osmotic effects mean that salts can only be excreted in solution – drinking these will make you more thirsty, hastening dehydration.

So while the nitrogen in urine makes it a great fertiliser and some insects will probe deposits for moisture and minerals, such drinking is best left to them. More effective in hot, dry situations is to soak an absorbent item of clothing with your urine and wear it as a hat, so providing evaporative cooling and some shade.

If the quantity of urine you intend to consume is particularly large and dilute because you gulped the last of your water too quickly, then drinking a first pass from your kidneys might conceivably help. The most efficient way to consume it is to drink about a mugful each time you start to feel thirsty. Then it won't enter the blood faster than the tissues can extract it.

Len Winokur
Leeds, West Yorkshire, UK

Organic lingerie

I am a midwife and a mother, and recommend using cabbage leaves for swollen, painful breastfeeding breasts, milk suppression and mastitis. I tell affected women to line their bras with the cold leaves. It seems to work, but does anybody know why?

Cate Turner
Kendall, New South Wales, Australia

Cabbage leaves are part of European folk medicine and have been described as a poor man's poultice (see www.bmj.com/cgi/content/full/327/7412/451-c). If there are controlled trials of the healing power of cabbages, they are not easy to find – Ed.

Cold cabbage leaves will have the simple effect of a cold compress and reduce heat in the same way as a cold flannel might (but without the drips). However, the beneficial effects of the cabbage are increased if you heat the leaves, by running a hot iron over them or by blanching in boiling water, before applying. The heat releases various anti-inflammatory chemicals as well as phytohormones. Leaving the leaves in the bra will have a slow-release effect as the body warms them and draws out beneficial chemicals.

Hot cabbage poultices have also been used for sprains and strains and to draw out splinters. I used the above remedy to treat a breast abscess (a side-effect of mastitis) resistant to antibiotics. Greek women used vine leaves for the same purpose. It would be interesting to find out if the leaves have the same chemicals in them.

Vivienne Tuffnell
Lowestoft, Suffolk, UK

Cabbages are members of the Brassicaceae, a large and diverse plant family. Among many other chemicals, brassicas produce glucosinolate compounds, one of which, sinigrin (potassium myronate), gives rise to the pungent smell associated with cooking cabbage.

In the presence of water and the brassica enzyme myrosinase, sinigrin forms 'mustard oils', which are noted throughout history for their healing properties when applied as a poultice. Crushed or chopped leaves are applied externally as a counter-irritant to ease swellings and painful joints and to cleanse infections, and a warming sensation can be experienced in the skin. Mustard oils can lead to blistering, however, so must be used with caution.

Richard Eden
Consultant botanist
Southampton, Hampshire, UK

8 Troublesome transport

Cluster buster

It's a popular cliché to say that you wait ages for a bus, and then three turn up at once. But is there any truth to this? Or is it a false impression formed because we notice coincidences more than other events? If true, are there laws governing this behaviour, and are there any natural phenomena that obey the same principles?

Clare Redstone
London, UK

The technical term for several buses arriving at once is 'bunching', and the reason for it is quite simple.

If for some reason a bus is delayed by a few minutes, there will probably be more people waiting for it than on average. This is especially true when the frequency of buses on a particular route is high enough (one every 10 minutes, say) that passengers tend to arrive at stops randomly rather than according to the timetable.

Any late running will therefore increase the time a bus has to pause to pick up passengers at a stop, especially if the bus driver has to sell or validate tickets on entry. The late bus is therefore made slightly later still. This effect is compounded at each stop, causing more and more people to be waiting, delaying it even more.

Meanwhile, the next bus on that route is getting a pretty quick run because many of the passengers it would have picked up are on the late-running bus. Eventually it catches up the bus in front and, if it doesn't overtake it, we are left with two buses trundling along together, with the next service behind catching up on them too.

There are a couple of solutions to this. The most obvious and widely practised is to include 'timing points' along the route – stops where a bus is scheduled to wait for a few minutes before continuing. A late-running bus may ignore this wait and so make up a few minutes, but on the downside a bus that is on time will put passengers through an unnecessary delay whenever it reaches a timing point.

Another solution is to give a bus priority at traffic lights (or some other priority over other vehicles), but only if it is running late. This allows the bus to make up time, after which it no longer requires privileged treatment. While this helps public transport run to schedule, it too has its disadvantages, including the cost of implementation and the disruption to schemes designed to optimise overall traffic flow.

Dean Purkis
Eltham North, Victoria, Australia

Inevitably at some point on the route there will be an unusually large number of passengers waiting; after all, they do not arrive at stops at a constant rate. The first bus along will have to deal with the crowd as best it can, while the bus behind it will have a smaller load to pick up and can go faster. As their journeys progress, this disparity becomes more pronounced until eventually the bus behind catches up.

This is a mathematical inevitability, not a matter of friendly drivers attempting to travel in packs. For further information you may want to read *Why Do Buses Come in Threes? The hidden mathematics of everyday life* by Robert Eastaway and Jeremy Wyndham (Robson Books/Wiley, 2005) or the briefer explanation in my book *What Are The Chances? Voodoo deaths, office gossip and other adventures in probability* (Johns Hopkins University Press, 2002).

Bart Holland
New York City, US

Loo clue 1

While on a cruise ship in Spitsbergen, Norway, we had to use toilets which would only flush if the lid was closed to create a seal. Sometimes you had to hold the lid down to ensure the seal was intact. How did they work?

Melanie Green
Hemel Hempstead, Hertfordshire, UK

I have seen this type of toilet on yachts. When the lid is closed to form a seal, the waste is automatically pumped out. This causes a reduction in the air pressure inside the bowl. The reduced pressure draws in water through a separate pipe, which is used to flush the bowl clean.

If the seat is damaged the seal may not form correctly and the toilet won't work. Therefore it is essential for all yachts to carry a spare seat.

Doug Grigg
Cannonvale, Queensland, Australia

Loo clue 2

How do toilets on airliners work? They have incredible suction. Is the low external pressure outside the aircraft's fuselage used to create this? And why do they operate a few seconds after you press the flush button?

Lance Martel
Lima, Peru

On the aircraft I fly (a Brazilian-built Embraer 195 jet) the two toilets are flushed by suction. To generate the low pressure required, two methods exist, depending on the stage of flight. If someone flushes the toilet at high altitude, the external atmospheric pressure is used, and the difference between internal and external pressure forces the waste into a holding tank at the rear of the fuselage. If the toilet is used at low altitude or on the ground a suction pump mounted near the waste tank provides the pressure difference and that is why it may take a second or two to flush once the button has been pressed.

Rob Cheesman
Pilot and First Officer,
Belfast, UK

Aircraft lavatories cannot be opened to the environment outside the plane for at least two reasons. First, flushing the toilet at altitude would cause explosive decompression of the cabin and second, if waste was scattered from the sky it would turn to ice and become a danger to people and structures on the ground.

Conversely, waste water from the food and drink galleys and hand basins is dumped from drain masts that open to the outside. These are electrically heated but occasionally chunks of ice do fall from the sky following a malfunction.

Toilet waste, on the other hand, is moved by a vacuum through a waste line to a holding tank that is emptied on the ground at the airport after a flight. If there is not already sufficient vacuum, pressing the flush switch starts up a generator, which depressurises the waste line. This takes about a second to operate, during which time a rinse valve opens and then stays open for a further second. A small, measured amount of rinse water is used to clean the toilet bowl. Then, after this 2-second delay, a flush valve opens and stays open for a further 4 seconds to ensure the toilet bowl is empty. The change in pressure eventually moves the waste to the holding tank.

The process cannot be left to chance. The flush sequence is governed by software and if a malfunction occurs the toilet shuts down and a signal is sent to the cockpit. If the waste tank becomes inoperative or fills up, the crew will be forced to put down for repairs.

Terence Hollingworth
Blagnac, France

Hang on lads...

At the end of the original 1969 movie of The Italian Job, *the thieves are in a bus hanging over the edge of a cliff, with a stack of gold bars about to slide out of the back doors. All the protagonists crowd to the front of the bus to balance out the weight of the gold. Their problem is they can't get at the gold, because every time they try to edge down the bus to grab it, the bus tips again and the gold slides nearer the doors. The movie ends before the dilemma is resolved. Assuming no outside action is permissible, how could the thieves save the gold?*

Helen Morten
Birmingham, UK

Having so ingeniously stolen the gold in the first place, this final little problem would not have caused much consternation among the gang. They would strip out enough overhead internal wiring to fashion a stiff 'rope', lasso the pallet of gold and drag it to the front of the coach. Next, they would dismantle the pallet and use the main timbers to inch the coach back on to the road using a rowing motion.

Tony Holkham
Boncath, Pembrokeshire, UK

There are several ways to solve such a balance problem. If the thieves are armed they could shoot and puncture the bus's fuel tank to drain it; the tank is typically in the rear of a bus and thus at the same end as the gold. Shooting the tank would not ignite the fuel, and draining it would tip the balance toward the thieves. Alternatively, a more time-consuming approach would be to run the engine until the fuel runs out. Again this would tilt the balance toward the front of the bus.

Another approach would be to break the windscreen so the heaviest of the protagonists can crawl out of the bus and hang from the front bumper, increasing their distance from the fulcrum and perhaps creating enough leverage for the lightest of the gang to reach some bars. Once a few bars are transported to the front of the bus the balance would be shifted enough to move the rest of them.

One more idea is for the thieves to escape from the bus via the broken windscreen all at the same time. The bus would fall but they would all survive. A standard gold bar, which weighs about 12.4 kilograms, is not easily broken or washed away. If the gang hiked down the cliff the bars should be easy enough to find, even if they were not contained by the smashed bus. Presumably the thieves would know how many to search for.

Anthony Castaldo
San Antonio, Texas, US

In 2008, to promote awareness of science, the UK's Royal Society of Chemistry asked the same question, with the conditions that the solution took less than 30 minutes to complete and did not involve a helicopter.

The winning answer, by John Godwin of Surrey, was for the gang to first break and remove two large side windows just aft of the pivot point and let the glass fall outside to lose its weight, followed by breaking two windows over the two front axles, keeping the broken glass on board to keep its weight for balance. One of the gang would then climb out through the front broken windows (but rest his weight on the ground) and deflate the bus's front tyres, to reduce rocking movement about the pivot point. He would then drain the fuel tank, which was aft of the pivot point; this would change the balance enough to let another man get out and gather heavy rocks to load the front of the bus. Then they could unload the bus until a suitable vehicle passes, hijack it and carry the gold away.

However, using the amounts given in the movie, the gold's weight would have been 3,200 kilograms, almost the exact weight of the Harrington Legionnaire coach it was in. The structural changes at the rear that would be needed to support the three 670-kilogram Minis the coach originally carried, plus the 3,200 kilograms of gold, luggage and fuel, would mean the weight aft of the fulcrum the bus was rocking on would exceed the total weight of the bus, sadly crashing both bus and question into the bottom of the valley.

Robin Hill
Crynant, South Glamorgan, UK

You can read the winning entry of the Royal Society of Chemistry's competition here: prospect.rsc.org/blogs/rsc/in-pictures-italian-job-entries – Ed.

The simple answer is that they could not save the gold. This is because the situation depicted is absurd. When I taught physics I would use this scenario to illustrate the very high density of gold.

To begin with, the amount of gold depicted would far outweigh the men on the coach so that their edging back and forth in the bus would have had very little effect. A cube of pure gold just 16 centimetres on each side would weigh more than an average man.

Also, the weight of gold in the back would have catapulted the coach over the edge as soon as it went out of control. If it did teeter on the edge by some film-maker's miracle, the weight of gold would create a lot of friction between the palette it was standing on and the coach floor, preventing it from sliding unless the angle was catastrophically large.

Great film, though.

Dave Oldham
Kingsley, Northamptonshire, UK

? Plane thinking

What is the lowest speed at which a full-size fixed-wing aircraft can fly without stalling?

Chris Szymonski
Neenah, Wisconsin, US

Even when an aircraft is in straight level flight, it is not horizontal. There is an 'angle of attack' between the longitudinal axis (the gangway) of the aircraft and the airflow, so the leading edge of each wing is higher than the trailing edge. This means that the air is deflected downwards resulting in a reactive force with an upward component (lift) and a backward component (drag).

Forward motion is needed to generate lift. The faster the aircraft is travelling, the smaller the angle of attack needed to generate the lift required to counter the aircraft's weight. As the aircraft slows down, the required angle of attack must increase to maintain lift. But increasing the angle of attack also increases drag, which slows the aircraft further, reducing lift. The stall speed is the speed at which increasing the angle of attack can no longer be used to generate lift, so the aircraft descends.

Stall speed is slowest when the aircraft is in straight level flight. When an aircraft turns it has to bank and so generates less lift, leading to an increase in stall speed. Increasing the payload or pulling up from a dive also increases stall speed, as does increasing the drag when the aircraft is in a landing configuration with its undercarriage down. For example, the stall speed of a Boeing 737 in straight level flight is about 220 km/h, compared with about 295 km/h when coming in to land.

Air France 447 tragically crashed into the Atlantic on 1 June 2009. Even though all components of the data flight recorder were recovered on 2 May 2012, the crash remains unexplained. It is thought that icing of the pitot tubes, which calculate airspeed from the pressure of the outside air flow, generated erroneous airspeed data. According to a report published on 27 May 2012 by the French Bureau of Enquiry and Analysis for Civil Aviation Safety, this led the pilot to increase the angle of attack, stalling the aircraft. The mystery is that the aircraft stayed in a stall throughout the 3.5 minutes of descent from an altitude of 38,000 feet, with an angle of attack consistently around 35 degrees. Without the benefit of daylight the crew had to rely on instruments in which they appear to have lost faith.

Mike Follows
Willenhall, West Midlands, UK

There is a class of light aircraft known as microlights that stall at speeds slower than 43 km/h. Some specialist aircraft can fly slower, but at the expense of a poor top speed. This is because the main limit to minimum speed is the weight per unit area of wing surface; called the wing loading.

Trailing edge flaps, leading edge slats and slots can assist a little, but the complexity and weight of such devices may degrade top speed while only reducing stall speed by around 10 per cent and increasing costs considerably.

An extreme example would be the aircraft that won the Kremer prize for crewed flight after being pedalled across the English Channel at a cruise speed of less than 30 km/h.

John MacDonald
London, UK

As John says, on 12 June 1979, the Kremer prize-winning Gossamer Albatross completed the 35.8-kilometre crossing of the English Channel in 2 hours 49 minutes, achieving a top speed of 29 km/h and an average altitude of 1.5 metres. It was piloted by amateur cyclist Bryan Allen, whose leg power drove its large two-bladed propellor. Fantastic – Ed.

The lowest stalling speed I know for a powered light aircraft is 50 km/h for the Fieseler Fi 156 Storch, a German reconnaissance and training aircraft used in the Second World War. The Storch has a low wing loading and is fitted with full-span leading-edge slats, Fowler flaps and ailerons that droop with the flaps.

These all combine to improve airflow over the aircraft's wing and prevent airflow separation over the rear part of the wing. This contributes to its low stalling speed and allows the aircraft to take off from a runway only 45 metres long and land in just 18 metres.

Thanks to these high-lift devices, at full throttle a Storch

can be flown at less than its normal stalling speed. Its nose is well up, so the engine thrust is helping to support the aircraft. I've read that it could stay airborne at 35 km/h when flown in this fashion.

Martin Gregorie
Harlow, Essex, UK

The lowest speed at which a plane can remain airborne is stationary, as in Harrier jump jets during vertical take-off. Here, the lift is provided by Newtonian reactive force, but this is achievable only when planes are not fully laden. In conventional flight, where a wing's lower surface encounters the air ahead of it obliquely and the airflow leaves the wing's rear obliquely, the thrust required to provide sufficient downforce far exceeds engine capabilities.

At the other extreme, the lowest reported airspeed a Boeing 747 can achieve without dropping as dead weight is 154 km/h – if minimally laden on approach for landing. If approaching into a headwind, for every km/h the headwind increases, the plane's airspeed can drop by an equivalent amount and still stay airborne.

Len Winokur
Leeds, West Yorkshire, UK

Rotor imbalance

A friend of mine says he's seen a helicopter loop the loop at an air show. Is it really possible? And if it is, how?

Quentin Jones
Edinburgh, UK

In theory, most helicopters can perform a loop. With enough

speed and distance from the ground they can use their momentum to do a loop and overcome the downward force produced by the rotors. The main problem is that when the helicopter is upside down, the rotors tend to bump around and flex too much. To overcome this you can use a rigid rotor – although rigid rotors then cause other problems.

Remote-control helicopters can actually fly upside down. They do this by adjusting the angle of attack of their rotor blades (this is called collective pitch) so that they are in the opposite position to normal flight.

They can get away with it because the forces acting on a small helicopter are lower and the joints and mechanisms are normally less complicated.

Johneng
By email, no address supplied

During a properly flown loop centrifugal force is greater than gravity, even when the aircraft is inverted. The pilot is pushed into the seat throughout and experiences positive *g*-forces.

However, if the loop is too large or it is flown at too low a speed, the pilot falls away from the seat, restrained only by a harness, and experiences negative *g*-forces. In older helicopters the rotor blades are flexible and hinged, flapping up and down, but under positive *g*-forces they always flap upwards. Under negative *g*-forces the blades may bend downwards far enough to hit the tail of the helicopter, with fatal results. This is why loops are discouraged.

Modern military helicopters have stiffer, hingeless, rigid rotors, giving much greater agility. Even under negative *g*-forces the rotor blades remain a safe distance from the tail, allowing loops to be flown safely.

Richard Whybray
Omagh, Tyrone, UK

You can see helicopters performing rolls and loops for yourself by searching for 'helicopter loop the loop' at YouTube.com – Ed.

Wheel to wheel

Car tyres have complex, deep tread patterns, presumably to channel water away in the wet. Yet the tyres on my friend's large motorcycle have hardly any tread pattern, though they are as wide as a car's. Their surface, more curved than a car tyre's, seems lightly punctuated by shallow, elongated S-shapes with large smooth areas between. I would have thought a bike would need more grip on wet roads than a car, so why are the treads different?

Peter Wiley
Redruth, Cornwall, UK

A motorcycle, having two wheels, needs more traction on a wet road than a four-wheeled car does, not because it slides more freely but because the consequences of even a minor slide can be disastrous. However, what your correspondent is describing is a tyre that is designed for a high-performance sport bike that would seldom be ridden in the wet.

Sport bike tyres tend to be made from soft, sticky rubber compounds and have the minimum tread necessary to comply with legal requirements, sacrificing both tread life and wet-weather roadholding to maximise dry-road traction. Their ability to grip a dry road is staggering, but sport bike tyres may only last 3,000 kilometres. Riders may wear out a set in as little as a week or two of hard riding, or a single day at the track.

Tyres designed for touring motorcycles that will be ridden in all weathers have more car-tyre-like tread patterns and are made of harder, longer-wearing rubber. They don't have the extreme dry grip of a sport bike tyre designed to adhere even when leaning at 60 degrees, but they maintain safe levels of traction on a wet road and may last up to 25,000 kilometres.

Phil Stracchino
Gilford, New Hampshire, US

Cars sit flat on the ground and their tyres are similar to a flattened cylinder. This means the tyre makes contact with the road along a line across its rim. Motorcycles, however, have to bank to go round corners. Their tyres are shaped more like a ring doughnut, and at any instant only one point on the rim is in contact with the road.

Any water that is trapped between a car tyre and the road acts as an unwanted lubricant, and it cannot escape without travelling a certain distance sideways. It therefore makes sense for the tyre to have grooves into which the water can quickly escape. Provided that these grooves run diagonally rather than straight across, it is easy to ensure that the tyre's area of contact with the road stays constant as it rotates.

Water can escape easily from beneath a motorcycle tyre given the small area of contact, so grooves would not help. What's more, as the tyre rotates, the point of contact sweeps out a circle around the tyre's edge – a circle that moves from side to side as the motorcycle banks. Any treads would turn this circle into a cogwheel, increasing noise and vibration.

Motorcycle tyres are generally made to be much softer than car tyres so that they can spread out more and maximise what contact area there is. This also makes them less hard-wearing.

Alec Cawley,
Newbury, Berkshire, UK

No skidding

Your question on motorcycle-tyre tread patterns prompts me to pose another motorcycle-related query. I was watching motorcycle racing at the weekend and noticed that the bikes were taking corners while tilting at well over 45 degrees from the vertical, in fact probably by as much as 60 degrees. Most of the time the bikes managed this without sliding and crashing. How do the bikes lean so steeply and corner on what looks like the side of their tyres without sliding?

Chris Grant
Wakefield, West Yorkshire, UK

Turning a corner, a motorcycle is forced outwards by centrifugal force as well as downwards by the force of gravity. If the turn is taken with bike and rider too upright, centrifugal force flips the bike outwards and throws the rider off. If the bike leans too much, gravity makes it lie down and the tyres lose grip. It then slides out, with the rider usually sliding along behind.

With the bike leaning over at the best cornering angle, the combined forces push the mass of the bike out and down through the contact patches where the tyres touch the track.

Car tyres only need to work while upright. They have a square cross-section and there is tread only on the crown, not the sides. However, a motorcycle tyre has a rounded cross-section and the tread extends onto the sides of the tyre. This allows the tyres to continue to grip the track, even when the bike leans over in a corner.

Richard Whybray
Omagh, Tyrone, UK

Tyre strategy

Why can't you mix cross-ply and radial tyres on a car?

Peter Spurring
Teddington, Middlesex, UK

You could mix cross-ply (called bias-ply in the US) and radial tyres, but the steering may feel odd and the car would become unsafe at high speeds, especially around corners. This is because the tyres are designed to handle road stresses in different ways.

All tyres are reinforced with cord, made of nylon or another textile. A cross-ply tyre has its layers of cord laid at an angle, or bias, to the centre line of the tyre, creating a herring-bone pattern beneath the tread. When the car goes around a corner, the whole tyre, including the tread surface, leans outwards. This lessens the contact pressure with the road on the inside edge of the tyre, reducing traction.

A radial tyre's cord layers lie at right angles to the tyre centre line. Look at the wheel from the side and the cords run radially from its centre, which is how the tyre gets its name. The tyre also has belts, usually of steel, beneath the tread. In a corner, the radial's tread tends to stay flat on the road, while the body of the tyre flexes towards the outside of the bend. There is less lifting of the tread from the road, giving better traction.

When a car with all radials or all cross-ply tyres loses traction in a corner, all four tyres theoretically break loose at the same time, allowing you to control the slide with steering to prevent spinning. This is what racing-car drivers are doing when you see them sliding around a corner.

With radials on the front and cross-ply on the back, the back wheels will lose traction first, causing the rear of the car to spin and resulting in loss of control. With the tyre mounting reversed, the situation is worse: the front will break loose first, causing immediate loss of steering control. The car would veer off to the outside of the corner.

Rick Dieckmann
Fort Jones, California, US

Stream of consciousness

One often hears on the news of yet another search for the black box flight-data recorders from a missing aircraft. Why is this data not transmitted periodically to a satellite or ground station so that in the event of the unexplained loss of an aircraft, it would be readily available?

Peter Cole
Sark, Channel Islands

Early flight data was recorded on photographic film that had to be housed in a box into which light could not penetrate. This is the likely origin of the term 'black box' recorder.

The black box now comprises the flight-data recorder and the cockpit voice recorder. There are also calls for the addition of a cockpit image recorder, which would record the external readings of the instruments and therefore what the flight crew actually sees.

David Warren was the first to develop a prototype of a combined data and voice recorder in 1957. As a research scientist at the Aeronautical Research Laboratory (ARL) in Melbourne, Australia, he helped to investigate a series of fatal accidents involving the De Havilland DH106 Comet in 1953 and 1954.

He recognised that access to a recording of what had happened in the aircraft before the crashes would have been invaluable. He recalled seeing the first miniature recorder at a trade fair and this inspired his black box. ARL assigned him the resources to turn his prototype into an airworthy instrument.

The aviation community gave the black box a largely lukewarm reception at first, until the crash of a Fokker Friendship at Mackay in Queensland, Australia, in 1960. This prompted Australia to make the black box recorder compulsory, and other aviation authorities followed suit.

According to the Aviation Safety Network, about 2,300 commercial airliners have suffered a breach of their fuselage, known as a hull breach, since then. As well as crashes and mid-air collisions, airliners have been shot down, or been the target of terrorist bombs and hijackings. The average number of hull breaches is now about 30 a year.

Of course there are instances where the cause of a crash is still uncertain even when the black box is recovered, but investigators have failed to recover the black box in only ten hull-breach incidents, which equates to less than 0.5 per cent of crashes.

As well as recording flight data to a black box, limited data is transmitted. However, communication with ground stations is sometimes lost, and even transmission to satellites is not perfect.

Even when encrypted, there is always the worry that transmitted data could be hacked, redacted or lost before it reaches accident investigators. Airlines might object to the cost of continuously transmitting data, set against the tiny chance of crashing and then failing to recover the black box.

More importantly, to improve the chances of recovering black boxes at sea they could be designed to be buoyant or to transmit a signal to be picked up by hydrophones.

Mike Follows
Willenhall, West Midlands, UK

And we learned more about black boxes from another reader – Ed.

On the subject of black boxes, my understanding is that the phrase was not invented in connection with aviation but was coined by Norbert Wiener, the cybernetics pioneer. He used it to describe a unit that appears on a system flow chart and performs some specified function. It has certain inputs and certain outputs but the internal workings are not specified. It may be a piece of hardware or a piece of software.

For instance, one might have a black box that performs a coordinate conversion, or another mathematical operation. Or it might have physical inputs and outputs: one can imagine a black box with sunlight and carbon dioxide as inputs and oxygen and diamonds as outputs.

John Ponsonby
Wilmslow, Cheshire, UK

Cold rush

I have heard that submarines travel faster in colder water. Why?

Brendan Reilly
Dublin, Ireland

Density and viscosity do not vary significantly over the range of temperatures typical of seawater, so differences in drag on the submarine cannot account for this. The efficiency of the vessel's propeller has more to do with it.

A spinning propeller creates regions of high and low pressure. Where the pressure falls below the saturated vapour pressure for a dissolved gas, the gas comes out of solution and forms bubbles. When the bubbles collapse, they create noisy shock waves; this is akin to what happens in a kettle just before it boils. These bubbles interfere with the action of the propeller on the water.

In warm water this bubble formation, called cavitation, happens at lower propeller speeds than in cold water. This is because the gas is more soluble in cold water and its saturated vapour pressure is lower. So a submarine can run silently at higher speeds in colder water.

The submarine's engine should also be more efficient in colder water, though this may not be very significant in practice. All engines that convert heat energy into mechanical work exploit the temperature difference between hot and cold reservoirs. The efficiency is given by the temperature difference divided by the temperature of the hot reservoir. Other things being equal, the temperature difference and hence the efficiency will be greater when the seawater (the cold reservoir) is at a lower temperature.

Mike Follows
Willenhall, West Midlands, UK

Gotta new motor?

Why do all new cars smell the same? It is very distinctive and seems unchanged over decades and brands, but does it come from paint, plastic or something else? And if so, why is it the same across all cars? Or have the manufacturers bottled a fragrance which they secretly spray to seduce new car buyers?

William Coley
London, UK

The smell stems from small molecules called plasticisers, added to the plastics that make up a large proportion of a car's interior. Plasticisers spread throughout the plastic to which they are added, sitting in between the polymer chains so that they can slip more easily over each other. This makes the plastic more flexible and less brittle. However, it is relatively easy for the plasticisers to escape into the air, and in an enclosed environment they can build up to the point that we can easily smell them.

No discussion of the smell is complete without a mention of the writer William Gibson, who is famously fond of describing it in his work. Here's an example from his novel *Count Zero*: 'And then he was in the cockpit, breathing the new-car smell of long-chain monomers, the familiar scent of newly minted technology… '

The smell has a sinister side, however. Exposure to some common plasticisers such as DEHP has been linked to adverse effects on male reproductive health.

Chris Harris
Charfield, Gloucestershire, UK

New cars tend to smell the same because they all have roughly the same blend of plasticisers, flame retardants, lubricants and other substances evaporating off interior components and outgassing from dashboard trim, seating foam, upholstery and suchlike. Most of these substances have been in use, little changed, over several decades, and vary little from one make of car to another, although a more expensive car with leather trim and upholstery is likely to smell a little less of plasticisers and a little more of leather-tanning oils.

The 'new car smell' has actually been synthesised and is available in aerosol cans. It used to be something of a trade secret among car dealers, who would routinely spray the interiors of used cars to make them smell like new. These days, the spray is advertised for sale in car magazines, and can even be bought over the counter at auto parts stores.

Why people would want to spray their cars is mystifying

to me. Surely you can't fool yourself into thinking that your car is newer than it really is?

Phil Stracchino
Gilford, New Hampshire, US

Wet wind

I have been told that the wind has more force on a yacht's sail in conditions of high humidity, such as in the tropics, because the higher water content of the air increases its mass, and therefore the force, on the sail at any given wind speed. Is this true? If so, is there an equation I can use to calculate the increased force?

Mike Stovold
London, UK

In fact, humid air exerts less wind force than dry air. At high speeds, the force exerted by a moving fluid is proportional to its density. There is a popular misconception that humid air is denser than dry air – probably because we feel more lethargic and tend to describe humid weather as 'heavy' and oppressive. However, the opposite is actually the case.

A water molecule, H_2O, has a molecular weight of 18. Air consists of approximately 79 per cent nitrogen, N_2, of molecular weight 28, and 21 per cent oxygen, O_2, with a molecular weight of 32. So air's effective average molecular weight is 28.84, much greater than water's 18.

From the ideal gas law, at a given pressure and temperature the number of molecules of gas in a given volume is constant, so if in humid air some of those molecules are water molecules, with their lower molecular weight, then humid air must have a lower density.

In fact, this lower density helps to drive the water cycle: the buoyant humid air rises to an altitude where it is cold enough to cause the water vapour to condense and fall back to earth as rain.

Although we might think of a gaseous water molecule as squeezing in between the air molecules and hence making the air denser, its actual effect is to force the air molecules apart to make space for itself. Only if the air were in a closed vessel would the evaporation of water cause it to become more dense, but then the pressure would rise too.

Conversely, it is easy to demonstrate how the volume and pressure drops when the water condenses back out again by shaking some warm water in a partly filled plastic bottle and then sealing the lid. As the water vapour cools and condenses, it will cause the air pressure to drop and the bottle will buckle inwards.

Of course, if the humid air contains fine droplets of water – as opposed to water vapour – then these will exert a large force as they impact on the sail.

Simon Iveson
School of Engineering
University of Newcastle
New South Wales, Australia

Air density decreases with rising humidity at typical sailing temperatures. At 25 °C, air at 90 per cent relative humidity is about half a per cent less dense than at 30 per cent relative humidity. However, the effect of air density on a sailing boat is complex because density causes drag as well as lift, depending upon the boat's 'windage' and sail plan.

Experiments using a velocity-prediction program developed for America's Cup yachts suggest that humidity has a negligible overall effect upon boat speed for a modern America's Cup-style yacht. (The software balances aerodynamic and hydrodynamic forces, which are derived from computational fluid dynamics models.)

Nevertheless, the effect of temperature and barometric pressure on air density is significant. A boat sailing in 15 knots of wind during a hot, humid, low-pressure day off Valencia, Spain, could be 1 per cent faster than sailing on a cold, dry, high-pressure day in Auckland, New Zealand, due to decreased drag caused by 10 per cent less dense air.

It's possible, though, that differences in wind shear and gradient between the locations would obscure this effect.

Christopher Miller
Kawau Island, New Zealand

It is possible that in some parts of the world there is a correlation between humid conditions and high barometric pressure, which would cause an increase in air density and hence a greater sail force for a given wind velocity.

High humidity may also cause the sail to become less 'leaky' than in dry conditions, either by causing the fibres in the sail fabric to swell or, less likely, by the condensation of water onto the sail, partially blocking the pores in the fabric.

Ian Vickers
Harrison, ACT, Australia

Whistle-stop

I recently bought a roof rack for my car and to my delight discovered it likes to whistle when my car approaches 100 kilometres per hour. But why does the whistling stop whenever I go under a bridge?

Peter Morris
Birmingham, UK

To produce a noise, the roof rack needs to be struck at a

suitable angle by a sufficiently strong, stable wind. The same effect comes into play when you blow across the mouth of a bottle. Taut strings also do interesting things in winds at various angles.

When your car travels at speed in an open space, a stable slipstream develops, which sustains your roof rack's whistle. A bridge, especially one without much space beneath, disrupts this flow around your car and redirects the wind. If this changes the angle at which the slipstream meets the rack then the whistle stops, at least temporarily.

To pin down this behaviour, you can easily experiment with roof racks of different sizes and shapes and drive your car in a range of environments. To fresh winds and bridges new.

Jon Richfield
Somerset West, South Africa

Cool mist

I learned at school that warm air holds more water vapour than cold. So why does my car's ventilation system demist the windscreen more quickly when I turn on the air cooling?

Chris Hills
Teversham, Cambridgeshire, UK

Air cooled by the car's air-cooling system cannot hold as much water as it did when it entered the system, so the excess condenses in the air-conditioning unit and runs out of the bottom of the car. The dryness of the air makes it more effective at clearing the windscreen.

Malin Dixon
Nuneaton, Warwickshire, UK

Warm air can hold more water vapour than cool air. For example, 1 kilogram of air at 35 °C can hold about 162 grams of water, while air at 10 °C can hold only about 86 grams of water.

Condensation on the inside of a car windscreen occurs when warm, humid air inside the car meets the cool windscreen. The temperature of the air on the windscreen drops, the air can no longer hold all the water vapour it is carrying, and condensation forms.

To demist the windscreen, the condensed water must evaporate – or be wiped – off the windscreen. The more water vapour air contains, the higher the water vapour pressure, and one of the main factors determining whether water evaporates off a surface is the water vapour pressure gradient. Evaporation increases the water vapour pressure of the adjacent air, and if the water vapour pressure of this air reaches the same level as its surroundings, evaporation may cease.

When cool, dry air is circulated by the car's ventilation system, this point is rarely reached. The air sucked in by the air-conditioner becomes dry as it cools and this creates a favourable water vapour pressure gradient for evaporation from the windscreen. The windscreen clears more quickly when the fan is on high speed as the air near the windscreen is changed more frequently.

Peter Kamerman
University of the Witwatersrand
Parktown, South Africa

Here's the final word, which is not just a matter of semantics – Ed.

The belief that warm air holds more water (the 'windbag argument') was disproved in 1802 when John Dalton found that water vapour pressure is almost the same in a vacuum as it is at normal ambient air pressures.

In other words, water vapour acts independently of the other gases in air. With an increase in temperature, evaporation increases, leading to more vapour and a higher vapour pressure.

At equilibrium, the evaporation rate and the condensation rate are the same. Rising temperatures increase the evaporation rate; falling temperatures increase the condensation rate (see Craig F. Bohren's *Clouds in a Glass of Beer*, published by John Wiley and Sons, 1987).

Steuart Campbell
Edinburgh, UK

Grip trip

Road surfaces on approaches to pedestrian crossings in the UK have a different composition, presumably to allow cars to stop quickly. What is this surface and how does it work? Why aren't all roads constructed using it?

Ron Harte
London, UK

The road surfacing referred to is a thin layer of calcined bauxite, or chippings of such fixed to the base of the road by epoxy resin. The calcined bauxite retains its roughness far longer than normal road surfacing, which eventually becomes shiny and polished by vehicle tyres. This layer allows vehicles to decelerate and stop more quickly without skidding. In the late 1960s, Shell marketed the earliest examples of calcined bauxite as Shellgrip.

In 1970, I was a graduate working with the Greater London Council (GLC). One of my tasks was to investigate the effectiveness of this treatment, which the council had installed at a number of sites with high accident levels. Because the treatment was significantly more expensive than normal resurfacing, the council naturally wished to know whether the additional expense was justified.

By comparing accident records for the three years before the surfacing was installed with those in the years after, I demonstrated that the treatment was a highly effective way of reducing accidents resulting in injury, as well as the more common damage-only, rear-end shunts between vehicles that frequently happen at pedestrian crossings and junctions with traffic lights. The potential saving was so great that it justified applying the treatment to virtually all similar sites, which the GLC did over the following years.

Your correspondent wonders why the treatment is not applied universally. The answer is mainly that the cost would be prohibitive. Also, if the whole network was surfaced this way, vehicle tyres would undoubtedly wear out much faster because of the higher friction created between the tyres and the road surface.

Roger Waddington
Banwell, Somerset, UK

Your questioner is referring to the high-friction calcined bauxite material that is applied by hand to the road surface in areas where anti-skid properties are a priority, such as at pedestrian crossings or on sharp bends, where the accident rate is high.

It works because the high-friction material offers greater grip for a car's tyres. There are three reasons why this material isn't used everywhere. First, all main road surfaces are checked for adequate skid resistance by local authorities on a routine basis, and as conventional road design normally meets skid resistance specifications without the application of a high-friction dressing, its use is restricted to areas of high stress.

Secondly, its use at pedestrian crossings serves another purpose – the different colouring provides a visual reminder to the driver of a hazard ahead.

Finally, the downside of this material is that road authorities have to maintain it, as it can wear off in patches, causing a hazard that requires remedial action at further cost, so its use is restricted to areas justifying the additional expense.

Alistair Donald
Service Manager (Construction Operations)
Transportation & Environmental Services department
Fife Council
Glenrothes, Fife, UK

9 Sporting life

Bounce back

I have a question about Geoff Hurst's famous goal in the final of the 1966 football World Cup, and it's not whether it crossed the line. His shot hit the goal's crossbar, deflected downwards, hit the ground and bounced out, away from the goal. I have seen similar shots since. Why does a ball that hits the underside of the crossbar nearly always bounce away from the goal after it hits the ground? The more powerful the shot, the more likely it is to do this.

Frank Horseman
Scarborough, North Yorkshire, UK

When the ball strikes the underside of the crossbar it acquires backspin. The crossbar exerts a turning force to the top half of the ball in the opposite direction to that in which the ball itself is moving. When the backspinning ball then hits the ground, the spin will tend to make it bounce back out of the goal and into the field of play.

This is easiest to understand if you imagine a backspinning ball dropping straight onto the ground in the goal area. The backspin would manifest itself as the ball rolling away from the goal.

The controversy surrounding Geoff Hurst's 1966 World Cup Final goal will likely be eclipsed, for a while at least, by that around Frank Lampard's 'goal' for England in their 2010 World Cup match against Germany in Bloemfontein, South Africa. Deflected by the crossbar, the ball landed behind the goal line but had enough backspin to bounce back to the underside of the crossbar. It then landed just outside the goal. This led the hapless officials to decide that the ball had never crossed the line.

Mike Follows
Willenhall, West Midlands, UK

Penalty tricks

Like many people I enjoy watching soccer, but am constantly amazed by how difficult some players or teams find penalty kicks. They frequently miss the target and don't even require the goalkeeper to save the shot. So is there anything science can tell us about taking a penalty kick? The goal is a lot bigger than the goalkeeper, so it should be easy to score. But so many penalty takers don't. Is there a well-founded, foolproof way of taking fear, emotion and human error out of the equation and guaranteeing a goal?

Peter Parker
Nottingham, UK

The Italian soccer team attributes part of its success in winning the World Cup in 2006 to the neurofeedback training team members had received.

Neurofeedback is a form of biofeedback – the process of becoming aware of bodily processes which normally occur unconsciously – and operant conditioning. By using electroencephalography to provide a flow of information to a person from their ongoing brain activity, enduring changes can be made in the person's brain-wave activity and accompanying behaviours.

As a very safe alternative to medication, neurofeedback

is used to treat attention-deficit disorder, epilepsy, brain injury and many other conditions. It is also a form of peak-performance training that has been used for a variety of tasks requiring calmness and focus under stress, such as stock-market trading, music performance and competing in many sports.

It's a relatively new field: the winter Olympics in 2010 in Vancouver, Canada, saw it used more widely than ever.

Laurence Lewis
New York City, US

The notion, mentioned by your previous correspondent, that the Italian soccer team used neurofeedback (consciously controlling bodily functions that normally operate subconsciously) to help win the 2006 World Cup does nothing to describe what their success was due to.

To answer whether there is a foolproof way to guarantee scoring a goal, there clearly is not. The question does not account for the presence of the goalkeeper, who can make the target for the kicker much smaller if they guess correctly which side the kicker is aiming at. That makes for more errors in kicking, even when the kicker aims where the goalkeeper isn't. However, game theory suggests that if the behaviour of either the kicker or goalkeeper is predictable, then an advantage can be gained – not a 'guaranteed goal', but an increase in the percentage of successful shots (or, indeed, successful saves).

To prevent this advantage, the only solution is for both kicker and goalkeeper to choose a side at random – not easily done in the human brain. And goalkeepers may learn to read small movements made by some kickers.

In order to be truly random, perhaps coaches could flip a coin and then send an order to the kicker, and do the same for the goalkeeper. In this case, the kicker needn't try to kick to a spot that the goalkeeper can't reach, it being understood that 50 per cent of kicks will likely be blocked and 50 per cent will get through.

So much for the theory. Even if each game were totally random, there would still be a winner in a penalty shoot-out, who might well have many beliefs as to the 'reason' for success. You might as well ask a lottery winner the 'secret' of their winning.

Don L. Jewett
University of California, San Francisco, US

Hands, knees, and...

In running races, the winner is the athlete whose torso crosses the line first. Hands, knees and heads don't count. How, then, does the device that records the winning times work? In particular, how does it know which part of the body has crossed the line first? And how does it distinguish between all the athletes, some of whom must be obscured from a beam across the finishing line by their rivals?

Denholm Pickering
Birmingham, UK

The 'beam' of the finishing line is really a camera that rapidly and repeatedly takes a picture of the plane of the finishing line. Over time, the slices of images of the finishing line are compiled into one picture and coordinated with the event clock to recreate a time-elapsed picture of the finishing line.

These pictures are how the times and places of the athletes are determined. They can be odd in appearance. For example, if an athlete's foot lands on the finish line itself, multiple photo slices give it the appearance of a ski.

While the picture itself is auto-generated, the times and places are determined by a human judge. For each athlete the judge finds the position of the athlete's chest. While this might not sound like a well-defined point, the athletes know how to contort their bodies at the last instant to make the chest cross the finish line at the earliest possible moment. Once the judge pinpoints the athlete's chest the corresponding time of crossing the finishing line is determined from the picture.

To minimise athletes blocking each other, the camera is raised and gets a bird's-eye view of the line.

I suspect there would be far more injuries from diving at the line if the hand rather than the chest were chosen to determine the winner.

June Andrews
San Francisco, California, US

Spin cycle

Why do track racing bicycles often have the rear wheel filled in with a solid disc, while the front wheel retains the standard spokes?

Brad Tyrer
Florence, Italy

Solid wheels are used because they provide an aerodynamic advantage over a spoked wheel. However, they also act like a sail and allow crosswinds to push against the bike. This cross force is acceptable for a rear wheel, but would interfere with the rider's ability to steer if it acted on the front wheel.

In addition, solid wheels are heavier than spoked wheels so do not allow for rapid acceleration. They tend to be used where the design goal is to help the rider maintain a steady speed – for example, in a pursuit event or time trial – but are not suited to a sprint race.

Steve Johnson
Eugene, Oregon, US

Golden balls

One criticism often levelled at soccer is that the game can be unexciting as not many goals are scored in most matches. What would be the likelihood of goals if the rules were changed to allow two footballs to be in play at the same time?

Tony Holkham
Petersfield, Hampshire, UK

If the game were modelled as a simple two-dimensional system, with the football ricocheting off players and sidewalls, then the number of times a ball crosses the goal line would be a function of the number of balls. For example, the number of goals scored would double with two balls in play. In practice, however, the number of goals might well increase by a bigger factor. When in possession of both balls, an attacking team could use one ball as a decoy, increasing the chances of scoring with the other.

But, quite apart from the difficulty of refereeing such games, putting more than one ball into play would increase the risk of injury as players collide in an attempt to play different balls. It would be much safer to experiment with the offside rule. For example, the rule could be changed so that only attacking players entering the penalty area before the ball would be considered offside. This would have the added benefit of preventing defences playing the offside trap, which can diminish the game as a spectacle. Or instead of having more than one ball, it might be entertaining to see more than two teams on the pitch.

Tri-Soccer consists of one ball but three teams, each defending its own goal, but permitted to score in either of the other goals. Along these lines the pitch could be circular with goals arranged 120 degrees from each other.

Mike Follows
Willenhall, West Midlands, UK

One of the attractions of soccer is that it is easily understood. There have been many ideas over the years for pepping up the game and increasing the number of goals scored. Changing the pitch layout, say, so three or four teams could play simultaneously would certainly lead to a higher-scoring and more exciting game. But having more than one ball on the pitch would increase the complexity – probably making the game unplayable and unwatchable.

If the only objective is to increase scoring then probably the best solution would be to increase the goal size. However, if the aim is also a more exciting game, the most effective changes are possibly those mooted by fairfootball.co.uk, a website dedicated to making the sport equally accessible to women playing alongside men.

John Harvey
Dinas Powys, South Glamorgan, UK

In the 1960s my family played Monopoly differently by renaming two Chance cards 'The Phantom Strikes Again!' When The Phantom appeared a new piece started at Go, took one turn around the board and then vanished – but any property it landed on went back to the bank, creating what now is just economic normality. My kids told me the neighbouring kids refused to play the game this way.

Val Sigstedt
Point Pleasant, Pennsylvania, US

Hobson's choice

If you are hit in the face by a cricket ball, would it be more painful at a distance of 50 centimetres from the thrower's hand, or 5 metres? To put it another way, at what point does the ball start decelerating?

Neil Christie
Adstock, Buckinghamshire, UK

Please, no experimenting with this unless you have access to a crash-test dummy – Ed.

The closer you are to the thrower, the more it will hurt. Any object moving through the air experiences an aerodynamic drag, so will slow down unless there is also a propulsive force acting on it. The cricket ball therefore starts to slow down the moment it is thrown. For the biggest bruise, you should stand so close that the ball hits your face just before it leaves the thrower's hand. (Whether this counts as a throw or a punch is up for discussion.)

Ben Craven
Menstrie, Clackmannanshire, UK

The ball starts decelerating as soon as the drag from air resistance exceeds the force exerted by the thrower – in other words, as soon as it leaves the thrower's hand. In theory, then, the pain increases the closer you are to the thrower.

However, it's worth checking this by doing a rough calculation. We know a cricket ball has a mass of about 0.16 kilograms and diameter of about 7.3 centimetres, and that in baseball, the pitcher can throw the ball at a speed of around 45 metres per second. With these numbers, plus standard values for the density and viscosity of air, we can work out the ball's Reynolds number – a measure of the extent to which it is affected by drag and turbulence. We can also calculate the deceleration force exerted by the air, which turns out to be around 0.5 newtons.

If we divide this result by the ball's mass, we get the actual deceleration, which is 3.1 metres per second per second. So in the tenth of a second or so that it would take the ball to travel a distance of 5 metres, its speed will have slowed by a tenth of that figure or just 0.3 m/s, to 44.7 m/s. This represents a loss in kinetic energy of just 1.3 per cent, so I don't think it will make a noticeable difference to the pain of the impact compared with standing 50 centimetres away. To properly compare the two scenarios in the question, you'd really have to experience them both – not something I would recommend.

Simon Iveson
School of Engineering
University of Newcastle, New South Wales, Australia

The cricket ball will trace out a parabola, and its path can be treated as the result of independent vertical and horizontal motion. Gravity influences only the vertical motion, and will have a much more significant effect on the ball's speed than the air drag.

If the ball has to travel upwards to its target because the recipient is higher up than the thrower, then gravity will reduce the vertical speed of the ball, making the impact less painful. The higher up the recipient is relative to the thrower, the more the ball will be slowed. Conversely, if the recipient is beneath the level of the thrower, gravity will accelerate the ball and make the impact hurt more. So the recipient is better off on higher ground, whatever the horizontal distance from the thrower.

If thrower and recipient are at the same level, there is no acceleration due to gravity to worry about. However, drag will cause the ball's horizontal speed to fall slightly as it travels. In this case, the recipient would be better off standing 5 metres away so the air can slow the ball as much as possible. This position also has the advantage of giving the longest time to duck...

Adam Coombes
The Cherwell School
Oxford, UK

The ball does not decelerate. In physics, any change in speed – whether an increase or a decrease – or a change in the direction of motion is known as acceleration. This may sound pedantic, but it is a point I, as a teacher of science, have to get over to my pupils. I know that children read *New Scientist*, so please, no more 'deceleration'.

Stephen Follows
Doncaster, South Yorkshire, UK

And in, and out...

I have attended several different types of exercise class over the years, and I'm always instructed to breathe in through the nose and out through the mouth while exercising. No instructor can tell me why. Is there a scientific reason for this advice?

Heather Shute
Oxfordshire, UK

Some rituals may produce benefits in the same way that a placebo can, but in principle the practice your questioner describes has some concrete advantages.

Gyms are often dusty or may be humidified largely by the exhalations of exercisers, who may also be sneezing out germs. In this environment, inhaling through your nose may be healthier than using your mouth, as the nose can filter out dust and fine droplets.

What's more, inhaling through your nose rather than your mouth makes it less likely that you will hyperventilate – and thus you are less likely to suffer the associated faintness and tingling brought on by a shortage of carbon dioxide in the blood. Air inhaled through the nose also tends to be warmer and moister than air entering via the mouth, so it does not dry or irritate the throat.

As for exhaled air, if it leaves via the nose it deposits heat in the nasal passages, retaining it within the body. Exhaling through the mouth is preferable as it increases evaporation, removing unwanted heat following exertion.

Jon Richfield
Somerset West, South Africa

At peak effort while exercising, you may well perform what is called a Valsalva manoeuvre. This also happens during evacuation of the bowel and in childbirth, and involves closing the glottis and contracting the abdominal muscles. The increased pressure in the abdomen results in a reduced flow of blood back to the heart from the abdominal organs and legs. If this manoeuvre is sustained at sufficient pressure, fainting can occur. As it happens, breathing out through your mouth makes it much harder to do a Valsalva manoeuvre.

Additionally, breathing in through the nose moistens the incoming air and so avoids drying out the airways. On the other hand, you lose moisture faster if you breathe out through the mouth than when exhaling through the nose, and this could explain why people who tend to exhale with their mouth when they exercise also tend to need more fluids.

Don L. Jewett
Former Director of Physical Therapy
University of California
San Francisco, US

On the bounce

Last week I played soccer for the first time in years. The goalkeeper kicked the ball high (probably about 20 metres into the air) and a team-mate called to me to head it. I realise I am lacking in skill, but the power of the ball striking my head knocked me off my feet, bruised my forehead and left me with a dreadful headache. Yet professional footballers seem able to head higher and faster-moving balls with no apparent damage or pain. What velocities and forces are they dealing with, and why did the ball leave me stunned but not a professional player?

Alan Nicholas
Barrow-in-Furness, Cumbria, UK

Imagine you were able to head a stationary ball to the goal-keeper from where you were standing that day. The force your head would have to exert on the ball would be equal in magnitude to the force exerted by the keeper kicking the ball. Since the ball exerts an equal reaction force on your head, it would be as if the keeper had kicked you in the head. Consider yourself lucky to just end up with a bruise and a headache.

In practice you would, of course, not be trying to head the ball all the way back to the keeper, but there are other things to consider with a falling ball. Ignoring air resistance, a football dropping vertically from 20 metres would hit the ground at a speed of nearly 20 metres per second. If you could reverse the velocity of the ball by heading it, assuming the ball has a mass of 400 grams and is in contact with your head for one-hundredth of a second, the force you experience is around 1500 newtons – the weight of two adults.

When professional footballers head the ball, they generally apply a small force to deflect it, usually to steer it into the net. That does not require taking the full force of the ball on the head. But if that is the only option, they use their necks, backs and knees as shock absorbers. In so doing, they increase the time the ball takes to slow down. By Newton's second law – force equals mass times acceleration, or in this case deceleration – this reduces the ball's deceleration and thus the force. However, as you have found, it takes skill and experience to get the timing right.

Mike Follows
Willenhall, West Midlands, UK

Form is temporary

Twice at the weekend I went out running. I ran at the same time on both days and for the same distance. I slept well the night before both runs, ate similar meals, drank the same amount of liquid and felt generally fine on both days. On the first day I bounced along, running as well as I would expect. On the second day it felt like I was running through treacle; I was lethargic and it was dreadfully hard work. I haven't subsequently gone down with any illness, so why did this happen?

Patrick Parratt
London, UK

I row at school, and know exactly how your correspondent feels on his runs. What he is experiencing is the result of his previous exertions. When someone runs, rows, or does any other kind of endurance activity, lactic acid builds up in the muscles due to incomplete respiration resulting from the body's inability to get enough oxygen to the muscles. If an individual does not warm down properly after exercise then the lactic acid is not removed, leading to weakness in the muscles the following day.

Felix Chapman
Salisbury, Wiltshire, UK

Establishing proper controls for a biology experiment is not as easy as it looks. Even using the same organism is not enough, because yesterday's guinea pig is not today's guinea pig, and yesterday's runner is not today's.

For example, yesterday the runner did not have a day's run behind him. While the body's feedback to exercise is poorly understood, the problem does not sound like lactic acid or any obvious physiological damage. Instead, it is more probably a mental defence against unnecessary exertion.

Jon Richfield
Somerset West, South Africa

As an enthusiastic swimmer, I have kept records of the time it takes me to swim 2.5 kilometres on almost 500 occasions over the past three years. On almost all occasions I felt like I was pushing myself to within 5 per cent of my maximum ability, and a majority of the times would be within 20 seconds or so of each other. Every so often, however, maintaining the usual pace would be completely out of the question and I would end up with a time substantially slower than usual, without any obvious explanation such as poor sleeping, eating or the onset of an illness.

I noticed that there appeared to be a roughly monthly cycle, and that the poorest result often occurred a few days after a very good one. When I charted all the results together, I found that more than half of my best monthly times fell within three days of a 30-day cycle, and most of my worst monthly times fell shortly afterwards. A graph of the three years shows a repeating pattern of gradual improvement followed by a 'crash' of almost 30 seconds in my time over a period of about five days. Having only noticed this pattern after most of the data had been collected, I believe I can discount any psychological effect.

Since discovering this I have searched unsuccessfully

for an explanation, which I would anticipate is related to some fluctuation in hormones or other biorhythm. Indeed, I haven't even been able to find an acknowledgement that this is a known effect in athletes, but I know that if I was competing at an elite level, it would be pointless to enter a competition that coincides with the worst days of my 30-day cycle. Is anyone aware of a study that examines this effect?

Ian Bradford
Flemington, Victoria, Australia

? Round and round

Athletics tracks are always run anticlockwise. Does this favour particular runners? Races could surely be run either way, so why never clockwise?

Peter Hallberg
Stockholm, Sweden

A race is like an unfolding drama and, as the Olympic tradition was begun by people who read from left to right, it would have been more natural for spectators to see athletes passing in the same direction. Given that dignitaries would have been sitting close to the finish line, it appears that the ancient Greek hippodrome was designed for anticlockwise racing.

The convention was sustained by the Romans, but clockwise racing has had its devotees too. Races at the University of Oxford were run clockwise until 1948 and ceased even more recently at the University of Cambridge. Races at the modern Olympic games were run clockwise until 1906, when some countries complained that they had already adopted the anticlockwise convention.

Mike Follows
Willenhall, West Midlands, UK

A common misconception about the origin of the modern elliptical running track is that it was based on the running track at the original ancient Olympic games. In fact, the running track for the prestige foot race (the stade) in the main ancient Panhellenic athletic festivals was almost invariably straight.

What was more or less elliptical, however, was the *hippodromos*, or chariot-racing track, and the shape of this was copied by the Romans for the main racetrack in Rome, the circus maximus. This was roughly rectangular – it had two long straight sides, with semicircular ends.

Illustrations of four-horse chariot races from the Roman period show that races were run anticlockwise. The most likely reason for this was the way in which the horses were controlled. The charioteer needed to control the reins of the two outer horses of the four. It was very important to ensure that the length of the rein of the outer horse in particular was controlled as effectively as possible when the chariot rounded the turning post. Evidence from the ancient world suggests that there was a bias in favour of right-handedness, just as today, and an anticlockwise race direction would obviously favour the right-handed charioteer.

When the modern Olympic games were devised, longer races were introduced. The most economical way of creating a track suitable for a range of distances is to have a circular or elliptical one. Baron Pierre de Coubertin, the founder of the modern Olympics, fancied himself as a classical scholar and it is possible that he was influenced by ancient pictorial evidence into assuming that there was some aesthetic or traditional reason why races should be run anticlockwise.

Maya Davis
Brighton, UK

I'm a runner (of no great talent) whose body is slightly larger on the left side. My left leg is, therefore, slightly longer than my right, so I would appreciate it if running tracks were reversed. They could even alternate, say every 10 years. I'm sure that I would be faster in a clockwise direction. By the way, horse racetracks in Melbourne, Australia, are anticlockwise but in Sydney they're clockwise.

Terry L
By email, no address supplied

Motor races tend to be clockwise, except in the US where oval racing is prevalent and then they are anticlockwise. Strangely enough, the non-oval Interlagos circuit in my home city of São Paulo is anticlockwise. What this tells us, however, I have no idea.

Thomas Keaton
São Paulo, Brazil

Which way to go?

When an arrow is fired from a bow I would expect from Newton's third law of motion that there should be recoil pushing the bow backwards towards the archer. Yet references appear to contradict this, suggesting the bow wants to follow the arrow. Why?

Colin Watters
Molesworth, Cambridgeshire, UK

Your questioner is correct in his expectation that the recoil from firing an arrow should attempt to push the bow back towards the archer. And it does. When an arrow is fired from a bow the arrow gains momentum in the direction of flight and, because momentum is always conserved, the firing mechanism – the bow, the archer and even the earth on which the archer stands – must gain an equal amount of momentum in the opposite direction.

The reason for the perception that the recoil is in the direction of the flight of the arrow is that, as the bow accelerates the drawstring and the arrow forward, the tips of the bow also move forward. They keep moving forward until the drawstring becomes straight and the tips and the drawstring come to a relatively abrupt halt. At this time, the bow exerts a forward force on the arm of the archer holding the bow – in other words, a recoil in the direction of flight.

The recoil thus varies over time – a backwards recoil as the arrow and components of the bow accelerate forward, and a forward recoil as the components of the bow decelerate. Although the second component involves less change in momentum than the first, it occurs over a shorter time, so the forces are higher. This is probably why the forward component of the recoil is more readily perceived than the backward component.

Ian Vickers
Harrison, ACT, Australia

In the case of a bow, the archer is part of the entire system and not separate from it. If the archer were wearing skates and standing on ice, he or she would recoil in a direction opposite to the trajectory of the arrow.

A bow is traditionally made from wood, connected by a string. When the archer draws the bow, the tips bend back. When released, the tips of the bow relax back to their equilibrium position, in the same direction as the flight of the arrow.

Modern bows can be oriented vertically so that, when the bow is drawn, the tips are parallel with the ground and to each other. When the archer releases the string, more of the recoil happens in the vertical plane. This means the upward and downward recoil cancel out.

Mike Follows
Willenhall, West Midlands, UK

10 Best of the rest

Jersey of many colours

When I look at an electric light bulb through the fabric of my maroon, acrylic school jumper, why does it seem to have a rainbow halo around it?

Thomas O'Hare (aged 8)
London, UK

The threads in the acrylic jersey are acting in a similar way to raindrops or ice particles in the sky when we see a rainbow. Light from the electric bulb is refracted when it passes through the translucent threads of the jumper and is separated into its constituent colours just as light splits when it passes through a prism.

The rainbow effect is more noticeable towards the edge of the acrylic threads because these are the thinnest, most transparent parts and refraction occurs more obviously and favourably here.

You probably won't get the same effect if you view the light bulb through the fibres of a woollen or cotton garment. These are natural fibres and do not have the same translucent, light-separating properties as acrylic.

Francis Keigh
Liverpool, UK

In the round

I attended a concert at Sydney Opera House in Australia. The ceiling has doughnut-shaped structures hanging from it. I presume they somehow enhance the music reaching the listener because I have seen similar structures in the Royal Albert Hall in London. What are they for and how do they work?

Noah Jacob
Sutherland, New South Wales, Australia

The acoustics of the Sydney Opera House are known to be very complex. My father worked as a radio engineer at the Australian Broadcasting Corporation in Sydney and was responsible for performing 'shot tests' inside the concrete shells used to construct the building. A shot was fired from a gun similar to a starting pistol and the echoes measured with an array of microphones connected to electronic detectors. This was necessary to determine the frequency of the reflections and the time it took for them to return.

Under some circumstances, soloists could become confounded by the mixed reflections and would gradually slow down to the point where they actually stopped singing.

The fibreglass doughnuts hanging from the ceiling were chosen to break up the echoes to ensure performers were not distracted and to improve the sound heard by the audience.

Max Bancroft
Narellan Vale, New South Wales, Australia

Both of the venues your questioner mentions had poor acoustics that needed fixing. The best concert halls immerse listeners in music by evenly distributing the sound around the room and by having a good level of reverberation. The Grosser Musikvereinssaal in Vienna, Austria, is considered one of the world's best music halls because its elaborate

ornamentation reflects and scatters the sound waves so well.

Too much reverberation gives an echoey space, like the unadorned Royal Albert Hall. This was rectified by suspending sound absorbers from the ceiling, which look like flying saucers.

On the other hand, if there is too little reverberation a hall sounds dead and even an orchestra cannot fill the space with music because it is absorbed quickly by the surfaces.

The transparent doughnut-shaped 'clouds' in Sydney reflect and diffuse the music back down to the audience without impeding their view. Low- and high-frequency sound waves behave differently so it is important to control a wide range to provide a good balance. Low frequencies (which have long wavelengths) can get past the clouds, which are small compared to these wavelengths, while the hard acrylic surfaces are effective at reflecting high frequencies (with short wavelengths) without absorbing too much sound energy.

Air itself absorbs higher frequencies quickly, affecting larger rooms more. So I would infer that the Sydney clouds increase the audience's exposure to higher-frequency sound by reflecting it down before the lofty air space absorbs too much of it.

Designing a concert hall is a complex business: you have to worry about the shape and size of the room, the materials used, the air conditioning, and even how many people will attend.

The following will help you to find out more: *The Musician's Guide to Acoustics* by Murray Campbell and Clive Greated (OUP, 1994); the *Proceedings of the 20th International Symposium on Room Acoustics* at bit.ly/sfEAYG; and the history of the Royal Albert Hall at bit.ly/tOP4Hy.

Iain Longstaff
Linlithgow, West Lothian, UK

Sound retail

Is there any evidence that piped music in shops, pubs and similar establishments increases sales? I avoid such places even if that is inconvenient or involves extra cost.

D. G. Shotton
Havant, Hampshire, UK

Not only does it increase sales generally, but specific music can increase sales of specific items. In a 1997 study, psychologist Adrian North and colleagues played stereotypically French and German music on alternate days in front of a display of French and German wines (*Nature*, vol. 390, p. 132). French music led to French wines outselling German ones, whereas German music had the opposite effect. Yet a questionnaire suggested customers were unaware that the music had an effect on their product choices.

Jon Sutton
Leicester, UK

Though I personally share the questioner's distaste, I know from involvement in marketing that such music does indeed increase sales. Tests have shown that music related to a product can affect sales directly; for example, playing overtly French music in a supermarket will increase sales of French wine or cheese, especially if other promotional tactics draw attention to these products.

This is probably because the music creates a mental association with a product, although exactly why remains a mystery. It works with smells too – the piped smell of fresh, ground coffee or baking bread also increases sales.

The impact of background music is less clear, though many retailers feel that a library-like atmosphere of silence can be off-putting. So, despite the cost (in the UK, payments must be made to the Performing Rights Society), such music is widely played. Indeed, some retailers make a feature of it. For example, coffee chain Starbucks displays a notice of what is playing and has the CD for sale.

That said, I am sure much music is off-putting, particularly if it is very loud, prompting some people to avoid the stores it is played in.

Patrick Forsyth
Maldon, Essex, UK

The writer is the author of Marketing: A guide to the fundamentals, *published by* The Economist – *Ed.*

Slap it on (or not)

What evolutionary, or indeed any other, reasons are there why women tend to wear make-up and men in general do not?

Petra Kirk
Ipswich, Suffolk, UK

Make-up provides an example of what animal behaviourists call a super-stimulus – a larger-than-life version of a normal cue that elicits an enhanced response. Women's make-up augments the features men instinctively find attractive, so promoting stronger interest and arousal.

Lipstick makes the erogenous zone of the lips more prominent. Rouge accentuates blushing, which indicates arousal on the part of a woman. Pupil dilation is also a sign of sexual attraction, and women in the past would put atropine (extracted from the poisonous plant belladonna) into their eyes to achieve this. Pupil dilation is most apparent in blue and green eyes, while blues and greens are the most popular eye-shadows. They are more natural than, say, reds or purples.

Make-up is also used to conceal blemishes and project youth, indicative of fertility, which can be desirable in a mate. Larger eyes and lashes help project the childlike aspect of feminine faces, eliciting the same protective instinct that ensures care and security by the male for the female and any offspring.

The reasons women tend to wear make-up are largely cultural and historic – men traditionally make the first move, so women resort to more passive means to attract a mate's interest. Sexual politics has progressed, but the gender association is deeply ingrained. However, men do wear make-up in certain contexts in other cultures, notably in oriental and Asian marriage ceremonies just as the bride does, and in some primitive societies.

TV actors and presenters of both sexes wear make-up for a different reason. Studio lights make skin look abnormally pale through the cameras. Because a pale complexion can be a sign of illness, our instinct to shun this is unlikely to boost ratings, so some time in front of the mirror is necessary for people on TV.

Super-stimuli are well documented. Hungry herring gull chicks peck a red spot on the hen's beak to induce it to regurgitate food, but peck even more vigorously at a red pencil with three white bars. Matings between different butterfly species are common – and some produce viable offspring. When this happens the species usually have a similar courtship and wing pattern – perhaps the only difference is that one of them is larger or has pronounced features.

As the questioner implies, the explanation need not be to do with evolution, more to do with how human society functions. To use the word 'evolutionary' to presuppose that evolution must invoke survival and natural selection can make such questions doubly leading.

Len Winokur
Leeds, West Yorkshire, UK

Kaboom squared

I attended a fireworks display where some of the fireworks exploded into a square shape. How do they do that?

Justin Muller
Tokyo, Japan

The firework was probably an aerial shell fired from a mortar tube. Such shells are usually spherical and contain stars, or pellets of coloured pyrotechnic material, arranged around an explosive charge. This charge both ignites and propels the stars in all directions at the apogee of the shell's flight.

The stars are arranged in a square when the shell is being constructed, positioning them at slightly different distances from the centre. The propulsive force on each star will be the vector sum of all the explosive forces surrounding it; because each star is in a slightly different position, the force on each is also slightly different. Those placed nearer the centre of the shell will experience less outward force than those further from the centre – the furthest being the stars that form the corners of the square. If the stars are uniformly arranged against the shell wall they will form a spherical burst. Shapes such as hearts and smilies are made by arranging stars asymmetrically.

One potential problem is that shells tend to spin on their trajectory, so there is a possibility that spectators will see this kind of firework burst end-on. If this were to happen, the stars would appear as a line rather than a two-dimensional shape.

Tony Charity
Lowestoft, Suffolk, UK

Smoke without fire

When pieces of wood are thrown onto a fire they burn up almost entirely, but a ream of paper can be retrieved from a fire after it has burned down with the central sheets virtually intact and only charred edges. What protects the paper from being burned throughout?

Colm McKean
London, UK

I spend most of my days in the lab burning things so am qualified to answer this question.

Almost everyone who has got themselves a wood stove in the hope of saving some money and the planet decides it would be a really great idea to use it to turn all the waste paper that puzzlingly appears from nowhere into heat. Single sheets of paper ignite so easily, and don't we use screwed-up newspaper to light the wood in the first place? Then they discover that, maddeningly, and plainly contrary to all the laws of the universe, piles of paper can barely be persuaded to burn at all. There are three reasons for this.

Although wood and paper are both mainly cellulose, the paper has had volatile oils processed out. These oils have a very low flashpoint and are mainly responsible for keeping wood logs burning. Second, the structure of wood and the internal boiling of those volatile oils makes it fall apart into chunks as it burns, exposing fresh surfaces to oxygen. Paper piles don't do this, and instead form a charred outer layer which actually protects the inside of the pile from heat.

Most significant of all, much of today's paper is whitened using rutile titanium dioxide, one of the most fire-resistant substances known. Surface cellulose can burn away leaving a thin film of rutile which prevents oxygen reaching anything underneath.

Paper used for art prints has even more rutile in it, making it very difficult indeed to burn. This, it is possible to argue, accounts for the occasional 'miracles' of the magically unburned pictures of religious figures saved from fires. I was especially intrigued by a fire at a Mormon tabernacle in 2011 in which a valuable original painting was completely destroyed, but a print of Jesus survived.

Glyn Hughes
Solid Fuel Technology Institute Information Officer (soliftec.com)
Adlington, Lancashire, UK

Trees of green

On my teacher-training course we are handed stacks and stacks of papers. However, I am aware of the need for environmental sustainability. So I would like to know how many A4 sheets are in the average tree used for paper making.

Joanna Fernandes
Plymouth, Devon, UK

A 20-metre pine tree with a trunk 20 centimetres across would yield about 45,000 A4 sheets (about 90 reams). This assumes that the density for wood is about 600 kilograms per cubic metre and that 50 per cent of the pulp is lost once the water and lignin are removed. It also assumes that the 'density' of the required paper is 70 grams per square metre. An A4 sheet is 210 millimetres by 297 millimetres; each sheet therefore has a mass of 4.4 grams (70 grams × 0.210 metres × 0.297 metres).

However, only about a third of the raw material for paper comes from virgin wood. 'Residue' or leftovers – wood chips and scraps that would otherwise be dumped or incinerated

to generate energy – make up another third. The final third is recycled paper.

Only trees not suitable for construction and making furniture and the like are harvested to make paper. Chemical pulping dissolves the lignin to make the paper stronger and less prone to yellowing, but this reduces the yield available for making the paper.

Mike Follows
Willenhall, West Midlands, UK

Index

D